OFF
WITH HER
HEAD

OFF WITH HER HEAD

THREE THOUSAND YEARS OF DEMONIZING WOMEN IN POWER

ELEANOR HERMAN

wm

WILLIAM MORROW
An Imprint of HarperCollins*Publishers*

HarperCollins books may be purchased for educational, business, or sales promotional use. For information, please email the Special Markets Department at SPsales@harpercollins.com.

FIRST EDITION

Designed by Nancy Singer

Library of Congress Cataloging-in-Publication Data has been applied for.

ISBN 978-0-06-309567-0

22 23 24 25 26   LSC   10 9 8 7 6 5 4 3 2 1

For Leslie Harris

History is a pack of lies about events that never happened told by people who weren't there.

—George Santayana, philosopher, 1863–1952

CONTENTS

INTRODUCTION 1

CHAPTER 1 THE ROOT OF MISOGYNY 15

CHAPTER 2 HER OVERWEENING AMBITION 30

CHAPTER 3 WHY DOESN'T SHE DO SOMETHING
ABOUT HER HAIR? 65

CHAPTER 4 THE DANGERS OF FEMALE HORMONES 97

CHAPTER 5 THE ALARMING SHRILLNESS OF
HER VOICE 114

CHAPTER 6 THE MYSTERIOUS UNLIKABILITY OF
FEMALE CANDIDATES 133

CHAPTER 7 WHO IS TAKING CARE OF HER
HUSBAND AND CHILDREN? 157

CHAPTER 8 SHE'S A WITCH AND OTHER MONSTERS 181

CHAPTER 9 SHE'S A BITCH AND OTHER ANIMALS 196

CHAPTER 10 HER SEXUAL DEPRAVITY 204

CHAPTER 11 SHE'S A MURDERER 250

CHAPTER 12 ADDITIONAL TOOLS TO DIMINISH HER 277

CHAPTER 13 MISOGYNOIR: WHEN POWERFUL
PEOPLE ARE FEMALE AND BLACK   301

CHAPTER 14 OFF WITH HER HEAD!   323

CHAPTER 15 RIPPING UP THE MISOGYNIST'S
HANDBOOK   338

ACKNOWLEDGMENTS   357

BIBLIOGRAPHY   359

INTRODUCTION

One of the many questions that have often bothered me is why women have been, and still are, thought to be so inferior to men. It's easy to say it's unfair, but that's not enough for me; I'd really like to know the reason for this great injustice!

—Anne Frank, *The Diary of a Young Girl*, 1929–1945

One spring day in 2019, I sat down to read a book for pure pleasure, a rarity for me. Because I often read as many as a hundred books in order to write one, my TBRFF (To Be Read For Fun) stack is halfway up to the ceiling. On that day, I chose a book on the amazing life of a controversial, smart, powerful woman. Settling into the story, I was disturbed to see a calculated misogynistic hate campaign to take her down.

She's unlikable.

She's untrustworthy.

She's sexually depraved.

She's disgustingly ambitious.

She's a spendthrift.

She busts men's balls.

You might be thinking I was reading about Hillary Clinton. In fact, I was reading Stacy Schiff's lush, Pulitzer Prize–winning 2010 biography of Cleopatra. I noticed that in Cleopatra's rise and fall, the story of her power and Rome's horror that a woman should wield it, there were uncanny similarities with Hillary Clinton's trajectory through the 2016 election and beyond. Certainly, we can judge some of the political choices of both Cleopatra and Clinton negatively. But what I found was more than that. In each woman's story, I discovered organized smear operations churning out unfounded accusations of sexual improprieties and criticisms of her ambition, untrustworthiness, appearance, and unlikability, accusations rarely made about male leaders either in the first century BCE or today.

Wait a minute, I said to myself as my jaw dropped. *Has this same stuff really been going on for more than two thousand years?*

Longer than that, I found, when I delved into female pharaohs who lived many centuries before Cleopatra. More than *three thousand* years.

In between Hatshepsut, the female pharaoh who came to power in 1479 BCE, and Kamala Harris, countless other powerful women have been subjected to almost identical sexist takedowns. Byzantine empress Theodora. Anne Boleyn. Elizabeth I. Catherine de Medici. Marie Antoinette. Catherine the Great. More recently, there is Chancellor Angela Merkel of Germany. Prime Minister Theresa May of the UK. Prime Minister Julia Gillard of Australia. Prime

Minister Jacinda Ardern of New Zealand. Governor Gretchen Whitmer of Michigan. Senator Elizabeth Warren of Massachusetts. European Commission president Ursula von der Leyen. And many more.

The accusations rarely have anything to do with experience, political mistakes, policy, or platform. They are name-calling caricatures that create two-dimensional comic book villainesses. She's a whore. A lesbian. A nymphomaniac. Frigid. Or all of the above. She's treacherous. Decadent. Power-crazed. Frivolous. Her voice is shrill. She is phony, inauthentic, unlikable, *unpresidential.* She is a witch. A bitch. She's ugly. Dresses poorly. Her clothes cost too much. Her butt is big. Her hair is wrong. She is angry, nasty, hormonally imbalanced, and irrational. She is a bad woman, a bad wife, a bad mother. She's a sexy vixen whose wanton ways and feminine wiles destroy good men. She is the very essence of moral turpitude, demolishing everything she surveys as she strides through life in four-inch stilettos, cackling wildly.

There is, I discovered, a clear pattern of vilification across the millennia and throughout history to bring down powerful individuals suffering from chronic no-penis syndrome. It's as if for thousands of years somebody has been passing along an instruction manual. I call it the Misogynist's Handbook.

The handbook was crafted to enforce the Patriarchy, a concept so towering it must be capitalized. According to Cynthia Enloe in *The Curious Feminist*, "Patriarchy is the structural and ideological system that perpetuates the privileging of masculinity. . . . [L]egislatures, political parties, museums, newspapers, theater

companies, television networks, religious organizations, corpora-
tions, and courts . . . derive from the presumption that what is masculine is most deserving of reward, promotion, admiration, [and]
emulation."

Though no one knows for sure, it is likely that the Patriarchy
arose in the shrouded mists of unrecorded human history. Its enforcement arm, the handbook, probably did, too, and is therefore
one of those rare books written long before writing was invented.
For many thousands of years, the handbook has kept women in
line—to differing degrees. In ancient Egypt, for instance, men almost always ran the show, but women had rights: to own property,
operate businesses, initiate lawsuits, make contracts, and divorce
their husbands with their assets intact. Their contemporaries to the
north, the Athenians, locked well-to-do women up in harems.
Though this book focuses mainly on stories from "Western" countries, the Misogynist's Handbook has been wielded against women
in cultures around the world, exerting an inexorable, hypnotic,
and often unquestioned pull on humanity. Consider, for a moment, foot-binding in China. Harems in the Ottoman Empire and
many other cultures. Bride burning in India. The Taliban wrapping
women up in ugly blue bags, faceless creatures to be beaten if so
much as a fingertip emerges from their hot, heavy shrouds. Rape
everywhere, in every era.

Which prompts us to ask: Is misogyny in our DNA, perhaps
arising as a method of forcing women to stay home to take care of
the young, ensuring the continuation of the human race? Or is it
a worldwide social construct, passed down from one generation to
the other? Or did it start with the one and continue with the other?

Whatever the cause, I think most of us—male and female alike—will agree it is high time to destroy the handbook, to rip out its pages one by one, and burn them as we cheer. But in order to do so, we must first understand exactly what is on them. That is what this book is about.

I need to point out that most of us who enforce the handbook, support it, and obey it unquestioningly are not bad people. Most of us are probably very good people. We merely continue the traditions we learned from infancy on, as our parents did, and theirs, going back thousands of years. Most of us certainly don't mean to harm anyone. Indeed, many of those *being* harmed are blissfully unaware of it. The millennia-long triumph of misogyny is largely due to its invisibility.

Another important thing to understand about the Patriarchy is that it doesn't hate women in general. It is actually quite fond of those of us who keep within its proscribed bounds, where we will be loved and praised for our gracious acceptance of its rules and regulations. Just consider for a moment all those male politicians accused of treating women badly who defend themselves by pointing proudly to their lovely wives and daughters as uncontestable proof that they are not misogynists. Of course, the lovely wives and daughters, smiling rapturously as they cook him dinner, are serving him rather than competing with him. But what happens when a person in possession of a uterus runs against him in an election? Is she now a threat to be taken down by a man concerned about his virility?

Yes, she is probably a threat, and threats must be eradicated in the most ruthless way possible. The Patriarchy selectively punishes those women who challenge male power, who refuse to be silent, and who are insubordinate to the unwritten but well-understood rules. And it uses the Misogynist's Handbook, those dependable measures discussed one by one in the following chapters, to do so.

An intriguing attribute of the Patriarchy is the terror of castration, which, you will find, runs throughout this story. There is a strong connection between a powerful woman, a man's fury at having to compete with her, and his fear of a resulting psychological, intellectual, and/or physical impotence. Much of patriarchal rage against women not staying in their assigned lane is likely due to this fear, whether it rises to the level of conscious thought or lurks, unseen but powerful, in the depths below.

IMAGINE DONALD TRUMP AS A WOMAN

Clearly, it is not always misogynistic to criticize women in power. But how can we tell when criticism is justified and when it is a chapter in the Misogynist's Handbook?

Here are some useful clues. For one thing, are critics analyzing her actions, her experience, her speeches, promises she failed to keep, shady deals, disastrous consequences, downright lies? Or are they deriding her appearance, voice, and personal relationships? Another way to determine sexism is to examine the words used to describe her. Are they the vague, coded words for misogyny: *unlikable, phony, shrill, inauthentic, unpresidential*? Are they blatantly sexist terms such as *whore, witch*, and *bitch*?

Let us take a quite recent example, that of Kamala Harris.

With an impressive résumé including a law degree, positions as San Francisco District Attorney, California Attorney General, and US senator serving on the high-profile Select Committee on Intelligence, no sooner had she been named Joe Biden's vice president pick than memes started blooming on Facebook and Twitter about her dubious sexual background. *Joe and the Hoe.* (That's right, her online detractors couldn't even spell a two-letter word correctly, confusing a prostitute with a garden implement, and maybe one day we can do something about the educational system in this country.) Kamala's grinning, open-mouthed head performing oral sex on the *J* of Joe. A meme of a man who won a contest by eating seventy-five wieners and beneath his photo, one of Harris saying, "Hold my beer."

Within days, a billboard appeared in Salisbury, Massachusetts, proudly sponsored by Rob Roy Auto, that read: "Joe and the Hoe. Sniff and Blow Tour. 2020."

Another meme portrayed Harris as the Wicked Witch of the West, green-faced, with a pointed black hat, holding a broomstick as Dorothy's house goes flying through the Kansas air. Yet another showed her as Medusa with writhing green snakes for hair.

"A mad woman," Donald Trump said of Harris soon after the announcement of her as running mate. "Totally unlikable," "extraordinarily nasty," and "so angry." A Trump campaign fundraising email called her "the meanest" senator.

"She's phony," said Trump's reelection campaign manager.

"Kamala sounds like Marge Simpson," tweeted Jenna Ellis, a senior advisor to Trump.

Not one of these comments was related to her experience, her positions, her mistakes, or her ability to step into the job on Day One and hit the ground running. That's how you know they are lifted right out of the Misogynist's Handbook.

Another way to determine whether misogyny is involved is to switch the gender of the individual in question and see if the results feel strange. Let us examine, for instance, the following statements made in this book about powerful women, but change the pronouns and other descriptors to masculine ones. Do they seem off, to you? Odd? Laughable, even?

"Men who are sweet, cheery and nonconfrontational will be rewarded."

"He doesn't have the right sort of body to be on TV."

"He seems not loyal at all and very opportunistic."

"He should show a little modesty."

"Should a father of five children, including an infant with Down's syndrome, be running for the second highest office in the land? Are his priorities misplaced?"

"It's unclear how the birth of a grandchild will affect his choice to run in the next presidential election."

"He's too bitchy. Humility isn't one of his strong points, and I think that comes through."

"Unbelievable! In the same week he wore the same suit twice!"

"On what should have been one of the proudest days of his political career, he bungled it with a less than flattering haircut."

"He is a monster of selfish ambition."

"His voice makes you envy the deaf."

"There is too much gravity in his voice. He needs lighter and brighter tones that introduce more melodious qualities and should take singing lessons."

"He reminds me of an angry third-grade teacher."

"He is unlikable with his smug facial expressions."

"He reminds me of a scolding father, talking down to a child."

"He has an air of inauthenticity, which is a major problem at a time when plastic politicians just aren't connecting with voters."

"He should smile a lot more."

"When the opportunity came, his ambition made him take up the knife and plant it in his opponent's back."

"He's a creep, he's a warlock, he's turned over to evil. Look at his face. . . . All he needs is green skin."

And my personal favorite: "He launched his political career in the bedroom by sleeping with a powerful woman."

Now go back and read them with the gendered words reversed. They sound more normal when talking about women, right?

Another method to identify misogyny is to picture a well-known politician as belonging to the opposite sex and see where that takes you. For instance, imagine Donald Trump as a woman. Let's call her Donna. During the 2016 presidential election, Donna Trump said the exact same things as her male twin, Donald, did in real life. Orange-faced, sporting a fantastically cantilevered helmet of yellow hair, she hid her weight under baggy, navy-blue pant-suits. Bellowing from the podium, she was angry, boastful. Only *she* could save the country. She called people nasty names, made fun of handicapped reporters and Gold Star Families, and refused to turn over her income tax returns. She lied and/or exaggerated on a daily basis. She had been married three times and cheated on all three husbands. She bragged about grabbing unsuspecting men's penises. Would Donna Trump have been viewed as blunt, honest, and refreshing? Would *she* have won the election?

Now imagine Hillary Clinton as a man. Harry Clinton said and did the exact same things as Hillary. He had been a popular senator and secretary of state, with high approval ratings, though he did send emails from a private server, as had his predecessors. Would Harry have been harshly criticized for his body shape, his suits, his thick ankles, and his voice? Would Harry Clinton have been portrayed in a thousand Pinterest images as a witch, stirring a cauldron or riding a broomstick? Would he have been called a bitch on countless T-shirts? Would his thoughtful, circumspect answers to media questions have been seen as inauthenticity, secretiveness, and untrustworthiness? Would Harry have been accused of running a child sex ring under a Washington, DC, pizza parlor?

Would attendees at Trump's political rallies have shouted:

"Lock him up!"

"Put him in prison!"

"Hang him!"

Which brings us to the title of this book, a reference to the feral, unadulterated bloodlust for women who step outside their place, misogyny on murderous steroids. As Melinda Henneberger, *Roll Call*'s former editor in chief, said of Clinton during the 2016 campaign, "Supporters don't just want to defeat her, but they seem to want to see her hurt. Disagree with her, dislike her, vote against her, but to even talk about hanging her?"

Rebecca Traister, writing for *New York* magazine, was alarmed watching the 2016 Republican National Convention. "I was not the only person in the room to be reminded of 17th-century witch trials," she wrote, "the blustering magistrate and rowdy crowd condemning a woman to death for her crimes."

If Hillary Clinton had walked into the 2016 Republican convention, or any Trump rally, for that matter, perhaps the audience, hungry for blood, would have torn her limb from limb. We can picture them as feudal peasants waving pitchforks and torches as they drag her to the scaffold to chop off her head or to the pyre to set her alight. Because "Lock her up!" is, in fact, only a slightly more civilized version of "Off with her head!" and "Burn the witch!"

In May 2020, opponents of Michigan governor Gretchen Whitmer converged on the state capitol threatening to have her lynched, shot, and beheaded because she closed down businesses to prevent the spread of COVID-19. Five months later, the FBI arrested thirteen men, members of a group called the Wolverine Watchmen, who had planned to kidnap her from her Mackinac Island summer home, put her on trial for treason, and blow up her boat and the bridge near her house to hinder police response. "Grab the bitch," wrote one conspirator to another. And "Just cap her."

"I knew this job would be hard," Whitmer said after she learned of the plot, "but I never could have imagined anything like this."

On January 6, 2021, rioters invading the Capitol in Washington, DC, were aiming to hang Vice President Mike Pence for a specific action: counting the Electoral College votes and confirming the election of Joe Biden. But their virulent hostility toward Speaker of the House Nancy Pelosi was for a far more amorphous crime, that of being a left-leaning female with great power. Her counterpart in the Senate, for instance, Majority Leader Chuck

Schumer, though heartily disliked by political opponents, has generated far less visceral hatred and far fewer death threats.

Evidently, a quick execution by hanging would have been too
good for Pelosi. Rioter William McCall Calhoun Jr., a Georgia lawyer, posted on Facebook, "The first of us who got upstairs kicked
in Nancy Pelosi's office door and pushed down the hall towards her
inner sanctum, the mob howling with rage. Crazy Nancy probably
would have been torn into little pieces, but she was nowhere to be
seen."

Torn into little pieces.

How can such violent, medieval hatred still be with us? Why
is it that when a woman seeks or wields power, it still sets people
howling at the moon and baying for blood? Why do they become
slavering wolves, itching to sink their fangs into female flesh? And
why are so many of those slavering wolves women?

A 2016 study by the social intelligence company Brandwatch
analyzed nearly nineteen million public tweets and found that
52 percent of misogynistic tweets were posted by women. In 2014,
the cosmetics company Dove conducted a study on five million
negative tweets written about women's appearances and body image. Eighty percent of the writers were women.

Pat Schroeder, the Colorado congresswoman from 1973 to
1997, wrote in her 1998 autobiography, "It was depressing that
while I was pushing hard for women's rights, the first to criticize
my agenda were often other women."

Shirley Chisholm, the New York congresswoman who served
from 1969 to 1983 and the first Black female presidential candidate, wrote in her 1970 memoir, "Women are a majority of the

population, but they are treated like a minority group. The prejudice against them is so widespread that, paradoxically, most persons do not yet realize it exists. Indeed, most women do not realize it. They even accept being paid less for doing the same work as a man. They are as quick as any male to condemn a woman who ventures outside the limits of the role men have assigned to females: that of toy and drudge."

Working for change, she added, is particularly hard for women, "who are taught not to rebel from infancy, from the time they are first wrapped in pink blankets, the color of their caste. . . . Women have been persuaded of their own inferiority; too many of them believe the male fiction that they are emotional, illogical, unstable, inept with mechanical things, and lack leadership ability."

What's behind a *woman's* misogyny? Where is this bizarre self-gaslighting coming from? Well, for one thing, we are probably unaware of how we, as women, are being sexist when we diminish and ridicule other women for their appearance and voice or call them those abjectly unimaginative names of *ho* and *bitch*. After all, the unending stream of misogyny in the media, including social media, and in every aspect of ancient history, current affairs, and everyday family life has been hammered into our heads since birth. It's so normal to be misogynistic that most of us are simply unaware of it. *I* certainly was until I did this research. There are words I will never use to describe a woman again, no matter how mad I am at her.

For another thing, many of us lovely wives and daughters are comfortable nestled in our proper places within the Patriarchy, cherished, feeling safe and protected by a system we understand. Many of us feel as threatened by the upheavals of societal change as men do, even if these upheavals are to our ultimate benefit. Change is scary.

Moreover, most of us were raised to want to be good girls. If we smile, and serve, and look the way men want us to, and never complain when they interrupt us and mansplain us, we are rewarded and praised. We fit in. And when many of us see a woman who doesn't fit in, doesn't even *try* to fit in, it's disconcerting. It threatens the status quo we are comfortable with. She is breaking the rules, violating time-honored traditions. She isn't being caring, supportive, loving. She is selfish, domineering, threatening, a slut, a bitch.

Good girls don't want to be like her. More than that: good girls want to tear her down. And the virulence of our misogyny might just equal the depth of our jealousy. Why don't *we* have the courage, the talent, the brains, to do what she is doing? Could we ever dare to break our bonds and soar so high?

And maybe, just maybe, we're hoping to purify ourselves of our own dark ball-busting shadow lurking deep within.

CHAPTER I

THE ROOT OF MISOGYNY

From her comes all the race of womankind,
The deadly female race and tribe of wives
Who live with mortal men and bring them harm.

—Hesiod on the creation of the first
woman, *Theogony*, eighth century BCE

In an astonishing coincidence, the creation myths of both the Bible and ancient Greece, arguably the twin pillars of Western culture, attribute all the world's ills—death, war, plagues, tsunamis, dandruff, flat tires, acne, *everything*—to the woman. It's all *her* fault.

In the biblical Book of Genesis, Adam and Eve wandered around naked and innocent in the Garden of Eden until Eve, listening to a serpent, became *ambitious*. The serpent said that if she ate of the fruit of a particular tree—which God had forbidden them to eat—she would be like God. And, seriously, who wouldn't want to be like God instead of wandering around nude and aimless in a garden? So she ate some, and then she gave some to Adam, and he ate some.

And they suddenly realized they were butt naked with interlocking body parts and grabbed some fig leaves to cover themselves out of pure shame.

Promenading in the garden, God was startled to see them hiding from him wearing fig leaves. When God asked them how they knew they were naked—a sure sign they had eaten the forbidden fruit—Adam threw Eve under the bus and said, "She made me do it." And God tossed them both out of paradise and posted a terrifying angel waving a flaming sword at the entrance so they could never get back in. And so all of humanity was consigned to toil, pain, and suffering for eternity because of a woman's unnerving ambition to venture outside of her place.

Of course, Adam could have said, "No, Eve. I will not eat the apple that God forbade us to eat. And neither should you." But alas, the naïve, harmless fellow couldn't resist her, what with her seductive, manipulative ways.

In about 200 CE, the church father Tertullian thundered, "And do you not know that you are (each) an Eve? The sentence of God on this sex of yours lives in this age: the guilt must of necessity live too. *You* are the devils' gateway: *you* are the unsealer of that (forbidden) tree: *you* are the first deserter of the divine law: *you* are she who persuaded him whom the devil was not valiant enough to attack. *You* destroyed so easily God's image, man. On account of *your* desert—that is, death—even the *Son of God* had to die."

Wow. We even killed Jesus.

Since thousands of years of history have judged all women by the badly behaved Eve, we might also deduce that all men are bumbling idiots, too stupid to question a terrible idea. But somehow that part got lost in translation.

In the Greek myth of Pandora—which the poet Hesiod described in his *Works and Days* in about 700 BCE—Prometheus, a fire god, created men out of mud. Zeus, king of the gods, disliked the nasty creatures and insisted on keeping them in a lowly state. But Prometheus, seeing his creations shivering in the cold and eating raw meat, disobeyed Zeus by stealing fire from Mount Olympus and giving it to them. Zeus was furious at Prometheus's disobedience but couldn't take the fire back, so he punished him by chaining him to a rock and sending an eagle to peck out his liver every day, which magically grew back at night, in an unending cycle of agony. Then the vengeful god decided to compensate for Prometheus's inestimable gift by bestowing a truly gruesome punishment on man: he created woman.

The other gods gave her many gifts, including the "sly manners and the morals of a bitch," according to Hesiod, "a shameless mind and a deceitful nature," and "lies and crafty words." They handed her a jar (which in later translations transformed into a box) with strict instructions never to open it—knowing of course she wouldn't be able to help herself—and sent her to Prometheus's addlepated brother, Epimetheus. Now, Prometheus knew Zeus had been planning something devious to hurt men, and he warned Epimetheus to beware of Greek gods bearing gifts.

But Epimetheus took one look at the gorgeous Pandora and married her. According to plan, the curious bride opened the jar, and the many plagues Zeus had stashed inside flew out across the earth, causing all the pain and misery ever since, the result of the first woman's disobedience. No one, it seems, has ever shifted any blame onto the dim-witted Epimetheus, who ignored the warning and let his lust rule the day. He and Adam both get a boys-will-be-boys pass for their nitwittedness.

How strange it is that both God and god brought evil into the world by creating woman, according to the stories written by men. And that in both creation stories, women are created in a different way than men. God creates Adam out of the earth, but Eve is formed from Adam's rib. Prometheus creates man out of his goodwill; Zeus creates woman to destroy the world.

Interestingly, the *Alphabet of Ben Sirach*, a satirical Jewish work of about 1000 CE, transformed an ancient Middle Eastern spirit of chaos named Lilith into Adam's first wife, whom God created out of earth at the same time he did Adam ("male and female created He them," Genesis 1:27). Lilith demanded equality in all things, including sex. She refused to always be in the subservient position on the bottom. But Adam refused. "I will not lie below," he said. "I will not lie beneath you, but only on top. For you are fit only to be in the bottom position, while I am to be the superior one." Lilith responded, "We are equal to each other inasmuch as we were both created from the earth." Even though it wasn't like she could find another man—Adam was still the only one on earth—she promptly dumped him, flying away from the proto-misogynist to exhilarating freedom. God realized he had to fashion Adam's new mate from something different, something to render the woman more docile; he used the man's rib to create Eve (Genesis 2:21–22). But that didn't end up so well either.

Alas, the daughters of Eve and Pandora have continued their bad behavior ever since their destructive ancestresses ruined the world. In the Bible, Delilah was bribed by the Philistines to find the source of the miraculous strength of her lover, Samson, one of the ruling judges of Israel. She discovered it was his uncut hair, and while he slept, she gave him a crew cut. Jezebel persuaded her husband, King Ahab of Israel, to worship strange gods and murdered a guy to steal his vineyard. After painting her face—always a sign a woman is up to no good—she was thrown by her eunuchs out of an upstairs window. Down below, horses trampled her, and dogs devoured her body. Clearly, the sly hussy got what she deserved.

Helen of Troy left her battle-scarred husband and ran off with a young stud, causing a ten-year war and the collapse of an entire civilization. The sultry sirens sang sailors to their doom on jagged rocks. Circe the witch transformed shipwrecked men into pigs. In the story of King Arthur, Guinevere smashed the Round Table through her adultery with Sir Lancelot, ending the bright, brief glory of Camelot, sending the land into chaos and war. Fairy tales are inhabited by beautiful evil queens with jutting cheekbones and stone-cold faces. The humpbacked witch in the forest eats children. The wicked stepmother enslaves her stepchildren or sends them into the forest to die.

The same story line of *blame the woman* runs not only through myth and legend, but also through history. Cleopatra used her dangerous allure to unman poor Mark Antony, the beefy Roman general, who lay supine on a purple couch as she dropped grapes into his mouth when he should have been conquering new territory for Rome. Anne Boleyn wrapped Henry VIII—that compliant, easily manipulated fellow—around her witchy sixth finger, making him ditch his faithful wife and give his own finger to the pope. Marie Antoinette said "Let them eat cake" and bought a billion-dollar diamond necklace while the people of France starved. Wallis Warfield Simpson used her feminine wiles to make Edward VIII abandon his duty to the realm.

Why are we so eager to blame a woman rather than admit to the shortcomings of men? Why can't we agree that Mark Antony was a drunken, womanizing fool? That Henry VIII was a ruthless sociopath who never did a thing he didn't want to do? That Louis XVI was a weak king, caught in the lethal mix of economic disaster and political change spiced by climate cataclysm? That Edward VIII was a Nazi-loving dolt who never wanted to be king, was looking for a way out, and took his bride on a honeymoon trip to visit Adolf Hitler? Why is the trope of the evil woman so powerful that it's still with us today? Why do we so often still give appalling men a pass with "boys will be boys" and "it's only locker room talk" while we demonize, belittle, shame, ridicule, vilify, slander, and silence women?

Perhaps the explanation lies in a question: How did this start?

The short answer is: no one really knows. But it is logical to assume that misogyny, the ingrained and institutionalized prejudice against women, must have been around a good while before the earliest woman-hating stories were set down in the Hebrew Bible and the Greek myths. Perhaps it started at the dawn of humankind when men could swing a big club and bring mastodon meat back to the cave, while the women were collecting wild herbs and berries, and caring for children, the aged, and the sick. Clearly, men had superiority in terms of physical strength.

But women were the only ones who nourished life in their bodies and brought it forth, just as the earth brings forth grains, vegetables, and fruits. And indeed, the reason women worked with crops may not have been solely due to their physical limitations in hunting mammoths, but because they could conjure life from the earth, just as they conjured it from their own magical bodies during the miracle of childbirth.

The unceasing rhythm of women's bodies was magically in sync with the moon itself, the divine mistress of the tides. Ancient peoples around the world believed that supernatural forces were at work when girls first menstruated and when women gave birth. The earliest man-made sculptures, some going back thirty-five thousand years, are female statuettes with exaggerated breasts, thighs, butts, and reproductive parts. Called "Venus figurines," they have been found from Siberia to France, and were carved from mammoth tusks, antler bones, and rocks. While it is impossible to determine exactly what these statuettes meant to their creators, many archeologists believe they represented fertility goddesses worshipped by ancient people for their ability to bring forth life.

Around 3,500 BCE, some eight thousand years after the creation of the most recent Venus figurine yet discovered, the tiny Mediterranean island of Malta was the Jerusalem, Mecca, and Rome of its day, a place of religious pilgrimage, where the faithful came from near and far to worship enormous statues of pregnant women. One statuette, pointing to her swollen genitals, appears to be on the point of giving birth. Traces of ochre paint—representative of blood, perhaps menstrual, perhaps from the birthing process—coat some statuettes. Strange twists of clay found on temple floors resemble human embryos. Carved triangles on the walls, point-side down, symbolized a woman's pubic region. And the temples themselves, with their curving walls and rounded chambers, resemble wombs. Founded a thousand miles away and some 2,700 years later, the most famous oracle in the world, the Greek Delphi, took its name from the word *delphys*, which means *womb*.

In her book *Sexual Personae*, feminist academic Camille Paglia wrote, "Woman was an idol of belly-magic. She seemed to swell and give birth by her own law. From the beginning of time, woman has seemed an uncanny being. Man honored but feared her. She was the black maw that had spat him forth and would devour him anew. Men, bonding together, invented culture as a defense against female nature." Woman, she wrote, represents the "uncontrollable nearness of nature," "a malevolent moon that keeps breaking through our fog of hopeful sentiment."

It is possible that men suffered horribly from uterus envy. Moreover, beings so powerful threatened to throw an orderly society into wild discord if not properly contained. What would women do if they fully unleashed their powers? Create whirlwinds, droughts, and floods; ride cackling on lightning bolts; release demons from the underworld? (It's no mystery why, until 1979, hurricanes were all given female names in the US.) Women's terrifying, destructive power must, therefore, be curbed, caged, constricted, diminished, kept in its place. Otherwise:

She eats the apple.

She opens the box.

She unleashes her fury on the Gulf of Mexico.

All hell breaks loose, and we are doomed.

"There is good principle, which has created order, light, and man," wrote the sixth-century BCE Greek philosopher Pythagoras, "and bad principle, which has created chaos, darkness, and woman."

What we do know for sure is that over time, the Patriarchy tore down women's life-giving magic, as it had to if it was to gain absolute control. The fourth-century BCE Greek philosopher Aristotle believed that a uterus was a kind of soil—dirt, actually—in which the man planted his perfect and complete seed. A woman merely provided a nine-month lease for a warm rented room. In the *Oresteia*, the classical Greek trilogy by Aeschylus, the god Apollo argued that it was impossible for a man to kill his mother, since no one actually *had* a mother, just a father and an unrelated woman who provided a safe gestation location.

In the thirteenth century, Saint Thomas Aquinas, arguably the most influential theologian in the history of the Catholic Church, declared women to be "misbegotten men," stating that they were inferior by nature and therefore incapable of leadership. Defective women had no place in business, politics, or finance.

All pregnancies, it was thought, started off as male, nature attempting to replicate its own perfection. But at some point in about half of pregnancies, something went terribly wrong, an irremediable birth defect, and the fetus became female. According to popular medieval literature, if a woman squatted with her legs spread very far apart, her female organs would fall out, and she would become a man. Though, given what women had to put up with, if this were true, floors everywhere would have been littered with ovaries and fallopian tubes.

For millennia, the most resounding theme of the Patriarchy among Greeks, Romans, Christians, and pretty much everybody else has been that good women stay home and be quiet. The fifth-century BCE Athenian statesman Pericles wrote, "The greatest honor a woman can have is to be least spoken of in men's company, whether in praise or in criticism."

In the biblical book of 1 Corinthians 14:34–35, Paul wrote, "Women should remain silent in the churches. They are not allowed to speak, but must be in submission, as the law says. If they want to inquire about something, they should ask their own husbands at home; for it is disgraceful for a woman to speak in the church." And in 1 Timothy 2:11–13, he advised, "A woman should learn in quietness and full submission. I do not permit a woman to teach or to assume authority over a man; she must be quiet. For Adam was formed first, then Eve."

If woman botched the beginning of the world, she will also signify the end of it. Dressed in purple and covered with precious stones and pearls, the Whore of Babylon, drunk with the blood of the saints, will ride a seven-headed dragon through the sky, laughing maniacally, waving a golden chalice filled with the filth of her adulteries. In case there's any doubt as to who she is, she's proudly had her forehead tattooed with "The Mother of Harlots and Abominations of the Earth."

THE MONSTROUS REGIMENT OF WOMEN

As disturbing as it is for any woman to tiptoe beyond her accepted bounds, when women hold positions of great power—pharaoh, queen, or high-level politician—the reaction has often been (and still often is) a howl of outrage.

Sixteenth-century Europe saw an explosion of misogyny as France, England, and Scotland all landed in the hands of powerful queens. The French religious reformer John Calvin believed that the government of women was a "deviation from the original and proper order of nature, to be ranked no less than slavery."

In 1558, John Knox, the Scottish fire-and-brimstone theologian then residing in Geneva, turned his sputtering outrage into a pamphlet called *The First Blast of the Trumpet Against the Monstrous Regiment of Women*. Taking aim at two monstrous Catholic women—Mary I of England and Mary of Guise, regent of Scotland—he dipped his pen in venom and wrote, "To promote a woman to bear rule, superiority, dominion or empire above any realm, nation, or city is repugnant to nature, contumely to God, a thing most contrarious to his revealed will and approved ordinance, and finally it is the subversion of good order, of all equity and justice." (Poor John Knox had some backpedaling to do when Elizabeth I came to the throne months later and was not amused by the blasts of his trumpet. "I didn't mean *you*," he wrote her pathetically, or something like it. "You are Protestant." But she never forgave him and refused to deal with him even when he became a huge political force in Scotland.)

Knox's monstrous regiment ended badly, which must have

gratified him greatly. Mary I of England died soon after his pamphlet was published, deeply unpopular for having burned some three hundred Protestants as heretics. Mary of Guise followed her to the grave two years later, reviled in Scotland for bringing in a French Catholic army to combat Scottish Protestant forces. Her daughter Mary, Queen of Scots' reign was brief and tumultuous; she lost her throne after only a few years by marrying the murderer of her husband. In 1567, her people rebelled, and she was taken to Edinburgh a prisoner. Along the route, the crowds cried, "Burn the whore! Burn her! Burn her! She is not worthy to live! Kill her! Drown her!" It must have sounded rather like the 2016 Republican National Convention.

When a woman in power wields it unwisely, as in these three cases that occurred within a decade, there is a great delight in her train wreck, a deliciously satisfying *I told you so*. But the patriarchal reaction is even harsher when she wields her power wisely, disproving the age-old stories of female incompetence, making the Patriarchy wrong.

Such was the case of Elizabeth I, the most successful monarch in English history, a woman ruling alone. During her reign, her enemies gnashed their teeth at the situation, and her subjects cheered her. But after her death, English men realized they didn't want the late lamented queen to serve as an example to their wives and daughters. If women started emulating Elizabeth's strength and independence, who would marry them, have sex with them, birth their babies, and cook them dinner? She had single-handedly proven everybody wrong by showing that a menstruating, menopausal, or old woman could run the country better than any man.

And that was far more horrifying than if her reign had been a complete disaster, which they could all happily have blamed on her gender as they tsk-tsked and said *I told you so*.

When Elizabeth's successor, James I, became king in 1603, according to feminist historian Retha Warnicke, pent-up misogyny exploded. A new generation of men took to blasting their trumpets in sexist symphonies. In 1615, Joseph Swetnam published his popular pamphlet called *The Araignment of Lewde, Idle, Froward, and Unconstant Women*, which slammed females for the devilish deceitful tarts they were, followed by countless other such works, putting women firmly back in their place. Swetnam's book alone saw ten editions over the next twenty years.

Perhaps the most egregious case of male horror at successful female rule was that of Empress Catherine the Great of Russia. Over the course of her thirty-four-year reign, Catherine expanded her borders by 200,000 square miles and promoted the arts, literature, and education, creating the Russian Golden Age. But ask anyone today one thing they've heard about Catherine the Great, and they will probably say that she died while having sex with a horse. Which never happened. The empress died of a stroke in 1796 at the age of sixty-seven. But her astonishing success proved misogynistic assumptions wrong, and something had to be done to trash her reputation forever.

There have, of course, been striking exceptions to the rule of misogynistic destruction. In recent decades, we find a few powerful women who, for reasons of culture, timing, and personality, managed to deflect tactics from the handbook with their legacies mostly intact, despite serious political missteps. Interestingly, all of them lived outside the US: Golda Meir, prime minister of Israel from 1969 to 1974; Margaret Thatcher, the first female prime minister of Great Britain, in office from 1979 to 1990; Ellen Johnson Sirleaf, president of Liberia from 2006 to 2018; and Angela Merkel, chancellor of Germany from 2005 to 2021. It is no coincidence that all of these women are known as "Iron Ladies." (Which is itself a sexist term. Women can't be strong and decisive? And have you ever heard of "Iron Gentlemen"?)

Such cases of handbook-resistant powerful females, however, are few compared to all the Jezebels and Messalinas, the Cleopatras and Anne Boleyns, the Catherine de Medicis and Catherine the Greats, the Marie Antoinettes and Hillary Clintons, whose tarnished and undeserved reputations litter the history books, novels, TV shows, films, and our cultural consciousness. Packaged for posterity, these women exert an archetypal resonance across the millennia, patriarchal proof that non-male people have no business in positions of great power.

Looking at the most vilified women in mythology and history, we see that they all got above themselves. Eve was ambitious and told a man what to do. Pandora was disobedient. Delilah took down an Israelite judge. Jezebel interfered with Israel's religious practices. Helen left an unfulfilling marriage. Cleopatra dared to rule a nation and challenge Rome. Anne Boleyn introduced religious reform to England. Catherine de Medici kept a disintegrating nation together for twenty-seven years. Marie's shopping was easier to blame than Louis's incompetence. Hillary ran for president. Kamala became vice president.

What I discovered in my deep dive into misogyny past and present is that a combination of lies, hatred, and sexism—often initially bruited about by one or two political enemies—have been repeated so often over the centuries that they have become inseparable from historical fact and are usually accepted without question. Today, the proliferation of sexist lies is much worse than long ago because all you need to do is push a button on social media and the lies fly around the world, potentially reaching millions in moments.

It is high time we reexamine history's most loathsome villainesses—the murderous harlot queens—as well as modern female leaders who have been painted in slime with the same ancient sexist brush. In most cases, we will find that with the tiniest bit of investigation the towering soufflé of vilification falls flat. There is no there there. The notorious legacies of most powerful women both ancient and modern are simply not deserved.

CHAPTER 2

HER OVERWEENING AMBITION

It's hard to be a woman.
You must think like a man,
Act like a lady,
Look like a young girl,
And work like a horse.

—Anonymous

The purpose of the Misogynist's Handbook is to keep women in their place. That place has expanded over time and in most areas around the world today allows for far more opportunities than, say, a harem. Still, the Patriarchy sees a woman's ambition to venture beyond that culturally assigned place as dangerous, jeopardizing society as we know it, something to be stopped in its tracks.

Which is why, as far back as we can go in recorded history, ambitious women had to pretend they actually weren't. That they were merely trying to be helpful and dutiful, feminine qualities applauded by the Patriarchy and unpoliced by the handbook.

In 1479 BCE, twenty-eight-year-old Princess Hatshepsut became pharaoh of Egypt. Female pharaohs were rare but occurred from time to time when the incest-riddled royal house ran out of healthy adult sons. Hatshepsut was the daughter of Thutmose I and, according to tradition, married her half-brother, who became Thutmose II on the death of their father. They had no son, but Thutmose II sired a boy with another wife. Thutmose II's son, the future Thutmose III, was, therefore, both Hatshepsut's nephew and her stepson. When Thutmose II died, the child was only two years old. His aunt stepped in to rule for him. Interestingly, Hatshepsut felt that she had to explain why she was taking power and almost apologize for doing so.

In a temple engraving, she stated that her divine father, the god Amun, told her to become pharaoh as her toddler nephew was clearly in no position to run the country. (And how can you contradict a god?) She pointed out that her biological father, Thutmose I, introduced her to his nobles as his heir before his death. Everyone, she wrote, wanted her to be pharaoh, and between the chiseled hieroglyphs she indicated she never wanted power for *herself*. Why did she put this on the temple wall? Probably because she knew that a woman's political ambition would be perceived with something akin to outrage.

Three thousand years after Hatshepsut, Elizabeth Tudor found herself queen of England at age twenty-five when her childless half-sister, Mary I, died in 1558. "The burden that is fallen upon me makes me amazed," she said. It *fell* on the poor woman like a loose roof tile crashing onto her head as she strolled by. She certainly didn't ask for it because it is such a *burden*. She continued, "And yet, considering I am God's creature, ordained to obey his appointment, I will thereto yield, desiring from the bottom of my heart that I may have assistance of His grace to be the minister of His heavenly will in this office now committed to me." She will yield to fate, obey God's will, amazed that such a thing should *happen* to her.

While Hatshepsut and Elizabeth inherited their positions—alarming enough to the Patriarchy—a woman vying for the most powerful position in the country by running for office against men drives sexists berserk. When thirty-five-year-old Benazir Bhutto became prime minister of Pakistan in 1988, she claimed that she was only continuing the agenda of her late father, Prime Minister Zulfikar Ali Bhutto, a kind of inherited position of its own. Her autobiography, *Daughter of Destiny*, begins with the words, "I didn't choose this life; it chose me." She wrote, "Whatever my aims and agendas were, I never asked for power. . . . Other women on the subcontinent had picked up the political banners of their husbands, brothers, and fathers before me. The legacies of political families passing down through the women had become a South Asian tradition. . . . I just never thought it would happen to me."

Becoming prime minister just *happened* to her. Because of her family. Never mind that her sister, who *really* didn't want to be prime minister, refused to be drawn into politics, fled the country, and, oddly enough, never became prime minister.

Even some "Iron Ladies" felt compelled to pretend they had no great political ambitions. One of those subcontinental female leaders that Bhutto was referring to was Indira Gandhi. In 1966, the

Indian Congress asked forty-eight-year-old Gandhi, daughter of India's first prime minister, the late Jawaharlal Nehru, to step in as prime minister for a few months until they could organize another government. Gandhi's reluctance to take the job resulted in greater pressure to do so. She finally answered that she would be "guided by the wishes of the Congress and its President Kamaraj." Modest, humble, unambitious, and dignified, she would be easy to manipulate, the power brokers assumed, and would put up no resistance when they kicked her out of office. They were wrong. Though Gandhi kept up the pretense of reluctance until she cemented her power, she decided to keep it. Other than a three-year hiatus, she remained prime minister until her assassination in 1984.

Even Golda Meir, a woman not generally known for her reticence, evinced feminine hesitation when asked to serve as interim leader of Israel after the fatal heart attack of Prime Minister Levi Eshkol in 1969. Seventy-year-old Meir, who had been thinking about retirement, announced that she couldn't make up her mind and wanted to discuss the situation with her children and their spouses. According to Meir, they were all agreed that she "really had no choice but to say yes." It was another way of saying the gods had decided, the burden had fallen on her, and she would be ruled by the wishes of others. Like Gandhi, Meir portrayed herself as the reluctant candidate, disavowing any ambition of her own. And, like Gandhi, once she had the reins of government in her hands, she was unwilling to relinquish them.

There's a reason why savvy female politicians have, for thousands of years, publicly expressed reluctance to assume power. Sheryl Sandberg, chief operating officer of Facebook, wrote in her book *Lean In: Women, Work, and the Will to Lead*, "Aggressive and hard-charging women violate unwritten rules about acceptable social conduct. Men are continually applauded for being ambitious and powerful and successful, but women who display these same traits often pay a social penalty. Female accomplishments come at a cost."

In 2010, two researchers at Yale University—Victoria Brescoll and Tyler Okimoto—conducted an experiment to measure public reaction to power-seeking candidates. Respondents were given the biographies of a hypothetical male and female senator with equivalent experience and qualifications and no mention of political affiliation. In some biographies, the woman candidate was described as "one of the most ambitious politicians" in the state; in others, the man was described that way. Strangely, both male and female respondents were less likely to vote for an ambitious female candidate and more likely to vote for an ambitious male candidate.

The researchers delved into the reasons for the respondents' answers. Why did they dislike ambitious female candidates? Seeking power, they found, did not fit the mold of an acceptable woman, who should be warm, caring, and sympathetic. "The intention to gain power may signal to others that she is an aggressive and selfish woman who does not espouse feminine values," Okimoto wrote. "Some voters even felt more contempt and disgust toward women when they expressed an interest in power, like there was something 'wrong' or repulsive about their lack of feminine communality."

Perhaps the oddest finding in the Yale study was that ambitious female candidates caused respondents to feel "reactions of contempt, disdain, anger, irritation, disapproval, disgust, and revulsion." Strong female leaders cause "moral outrage" and "cognitive confusion," causing our brains to explode.

Someone was certainly angry at Pharaoh Hatshepsut, who, once firmly settled on the throne, never did turn over power to her nephew. In a peaceful and prosperous reign of more than twenty years, she expanded Egypt's borders, built new temples—including the stunning Deir el-Bahri at Luxor as her mortuary complex—increased trade, and enriched the nation. By the time she died at about the age of fifty—inadvertently poisoning herself with carcinogenic skin lotion, tests on her mummy revealed—her nephew was well into his twenties.

After her death, Thutmose III, finally pharaoh in more than name, chiseled her image and name off the monuments and gave credit for her accomplishments to her father and grandfather. At one point he evidently decided it was easier just to smash her statues to bits. Hatshepsut biographer Joyce Tyldesley believes that Thutmose III feared his aunt's resplendent reign could persuade "future generations of potentially strong female kings" to not "remain content with their traditional lot as wife, sister and eventual mother of a king" and take the throne. To prevent more such monstrous women grabbing for power, he literally erased her.

After Hatshepsut's reign, portraits of Egyptian queens—which had been equal in size to those of their spouses—shrank to Lilliputian dimensions. The queen was now a doll-like creature, impotent, pitiful, barely rising to the pharaoh's knee, clearly in no position to seize power from a man.

For thousands of years, royal women like Hatshepsut have been called upon to rule while the men were off waging war or young kings were growing up. These women were praised for stepping modestly from the shadows, dutifully wielding power for a few months or years, and lauded even more for shutting up and disappearing from public view when the male was ready to rule. Serving in a time of need is self-sacrificing, motherly, graced with the official stamp of patriarchal approval. Handing over the reins of power to the returning husband or grown-up son with a huge sigh of relief and disappearing from public view is what a good woman would do. Hatshepsut was clearly not a good woman.

In medieval and Renaissance Europe, a combination of factors—lack of male heirs, the accidents and fatal illnesses of kings in their prime—resulted in many royal women serving admirably as regents for younger brothers, sons, and nephews. As one early sixteenth-century French writer, Jean de Saint-Gelais, put it, the person of an underaged king should be placed in the hands "of those nearest to him who are not entitled to succeed." A male relative in the line of succession might be tempted to add a little secret sauce to his ward's pigeon pie. A female relative who could never inherit the throne would derive no benefit from harming him and, indeed, would lose what power she had.

Based in Spain, Habsburg emperor Charles V (1500–1558) ruled such a vast empire that he needed a trusted relative to rule over the northernmost portion, the Spanish Netherlands. Over a period of several decades, three women served as regents—Charles's aunt, Archduchess Margaret of Austria, then his sister Mary of Hungary, and finally his illegitimate daughter, Margaret of Parma—all of them skilled politicians who happily gave up power when asked.

Perhaps the most remarkable Renaissance regency was that of the French princess Anne de Beaujeu. Her father, Louis XI, granted her the powers of a regent after his death—though she never held the official title—until her younger brother, Charles VIII, was old enough to take charge. The old king admired Anne's political cunning, calling her "the least foolish of women," surely a great compliment at the time, though it doesn't sound like much now. When Louis died of an apoplexy in 1483, twenty-two-year-old Anne ruled France for her thirteen-year-old brother, an arrangement that lasted until he turned twenty-one in 1491. Called Madame la Grande, Anne was described by her contemporaries as "a woman truly superior to the female sex . . . who did not cede to the resolution and daring of a man." She would have been born to the height of sovereignty "had nature not begrudged her the appropriate sex."

Contemporaries often compared Anne to her late father, known as Louis the Prudent, a gifted monarch. But while these comparisons would have been complimentary to a prince, they come off as a bit harsh when describing a princess. One judged her "haughty, unrelenting, guided in all she did by her father's maxim and just like him in character." She was "a shrewd woman and a cunning if ever there was one," wrote Pierre de Bourdeille, abbé de Brantôme, in the following century in his *Book of the Illustrious Dames*, "the true image of King Louis, her father. . . . She held her grandeur terribly" and could be "quarrelsome." She was a "virago." (Historical note: these were fifteenth-century gender-coded words for what today we would describe as feisty, difficult, shrill, nasty, and bitchy.)

Anne's secretary Guillaume de Jaligny wrote that she stayed

always by young Charles and that nothing related to him or France was done without her knowledge and consent. Male courtiers resented her power, and some of them "claimed that the King was kept in subjection and his authority was usurped." Anne had clearly gotten beyond the proper boundaries for a woman, even for a king's daughter. "She wanted to hold the highest place and to govern in all things," wrote Brantôme. She was "so ambitious."

The young king's cousin, François II of Brittany, was furious that Charles was held in "subjection . . . *by a woman.*" Louis II, duc d'Orléans, the presumptive heir to the throne, told Charles, "Madame de Beaujeu, your sister, . . . wants to keep you in leading strings and to have rule over you and your kingdom." D'Orléans tried to seize the person of the king and proclaim himself regent; Anne had the guards clap him in irons. When her brother turned twenty-one, Anne stepped gracefully aside and returned to her lands in Bourbon.

"VIRTUOUS DEMEANOR, GODLY CONVERSATION, SOMBER COMMUNICATION AND INTEGRITY OF LIFE"

Conjure up an image of Anne Boleyn and you will likely envision a cruel, cunning woman of overweening ambition, using the promise of tantric sex to rip apart a royal marriage and trash the national religion to make herself queen of England. It's a compellingly misogynistic story: a bitchy vixen against a pious middle-aged wife, the king a hapless victim of his own lust and the little tramp's manipulation. And the ending—the villainess losing her head on the scaffold—is a satisfying morality tale. Just look what happens to women like that. Even if she didn't commit the crime she was executed for—adultery with five men, one of them her own brother—she certainly had it coming.

Henry VIII's second, most interesting wife exerts a compelling fascination across the centuries. She looms large in Western history and, as the poster child for misogynistic tactics used to take ambitious women down and shame them forever, it is worthwhile to spend some time delving into her story.

Let's start with a close examination of the major sources that historians have used to tell it. For starters, they were also her fiercest enemies. The first and primary source, Eustace Chapuys, Spain's ambassador to England, was devoted to Henry VIII's first wife, Catherine of Aragon, writing to his royal master every bit of unflattering gossip on Anne he could dig up at the English court. He called Anne "that whore" and "the concubine" and obsessively wrote the most outrageous rumors that somebody heard that somebody else had heard.

Fifty years after Anne's death, Catholic propagandist Nicholas Sander published a book during the reign of her Protestant daughter, Elizabeth I, whom he despised. This second source of Anne's life portrayed her as a heartless whore, with all the markings of a witch, who bore a monstrous, deformed fetus (a sure sign of involvement with Satan) and slept with pretty much anybody and everybody.

Other sources, however, provide a different portrait of Anne, one that refutes the image of her as a selfish, wanton temptress. Anne's chaplain, William Latymer, knew her well; he was the only biographer of Anne's personally acquainted with her at all. Her contemporary John Foxe wrote an account of her life in his wildly popular book on evangelical martyrs. Though he was clearly heavily biased in her favor, making her into a saint and martyr—which is admittedly going a tad far—historians have generally agreed that his facts were accurate. And George Wyatt, the grandson of Anne's admirer Thomas Wyatt, reported stories of Anne passed down in the family. But these tales of decorous behavior, zeal for religious reform, and charity to the poor aren't nearly as engaging as those of a middle-aged queen and her ambitious lady-in-waiting wrestling over the affections of a king.

Legend will tell you that Anne, a lady-in-waiting to Queen Catherine, plotted to drive the king crazy with unsated lust so he would divorce Catherine—who had only given him a girl child instead of the desired male heir—and make Anne queen. Examining the sources, we find that initially, at least, her sole ambition was for the besotted monarch to just leave her alone. For instance, in 1526 when the king and queen went off on a summer progress—a tour of the countryside to hunt and be seen by the people—taking a small group of courtiers with them, Anne used the opportunity to flee to her parents' home, Hever Castle, and didn't return to court for nearly a year, despite the king's repeated invitations. Misogynistic history affirms that the cunning trollop did so to send the king into a frenzy of frustrated desire. But this doesn't pass the laugh test. Numerous noble families were eagerly pushing nubile daughters toward the king's bed in the hopes of lands and honors. And "out of sight, out of mind" is usually what happens instead of "absence makes the heart grow fonder."

So why did Anne run away? George Wyatt claimed that Anne

fled the king for "the love she bare to the queen whom she served." It seems Anne cared about the woman she had spent so many years serving and was embarrassed that the king was stalking her. Also, the king had made Anne's sister Mary his mistress for a time and un-ceremoniously dumped her. Perhaps, too, Anne had never stopped loving Henry Percy, heir to the earl of Northumberland, whom she had sought to marry in 1523. But Percy's father evidently felt Anne wasn't good enough for his son and forced him to marry a wealthy heiress. If Anne became the king's mistress for a season, she would just prove Percy's father right.

By the summer of 1527, however, it seems King Henry and Anne had come to an agreement. The king would have the pope an-nul his marriage to Catherine on a technicality, and he would marry Anne. Henry duly set the annulment in motion. But two issues pre-vented Clement VII from granting the request. The most pressing was that he was a prisoner of Charles V, Catherine's nephew, who had just sacked Rome. The second was that, over time, the pope had learned that Anne was a zealous religious reformer, bent on purifying the Church of corruption and superstition—the selling of indulgences and venerating fake relics—and a supporter of prohib-ited books deemed heretical. Though Clement would remain fairly subservient to Charles V long after he departed Rome, the pope may have been more inclined to issue an annulment if Henry had declared his intention to marry a devout Catholic princess.

Anne's chaplain William Latymer wrote that when Anne dined with the king, she would energetically debate scripture. Anne intro-duced Henry to "heretical" books advocating Church reform. John Foxe wrote that she was a "zealous defender" of the gospel.

Which brings us to what might have been a crucial reason why Anne accepted the king's offer of marriage. Not only would it provide an honorable way out of his embarrassing pursuit of her, which was rapidly rendering marriage with a man of suitable rank and good family impossible, but, according to Hayley Nolan's eye-opening reevaluation of Anne's story, *Anne Boleyn: 500 Years of Lies*, perhaps Anne realized how much she could assist Church reform if she became queen. That's not to say she didn't want the power and glory that came along with the position (and what woman would have turned it down?). But it adds a totally new facet to her motives.

There was, though, the thorny problem of Queen Catherine. The king, having proposed to one woman, was still married to another, and living with her while awaiting the pope's decision on the annulment. On December 25, 1528, the French ambassador Jean du Bellay wrote, "The whole court has retired to Greenwich, where open house is kept both by the King and Queen. . . . Mademoiselle de Boulan [Boleyn] is there also, having her establishment separate, as, I imagine, she does not like to meet with the Queen."

Hell no. Meeting with the queen would have been terribly uncomfortable for anyone with an ounce of integrity. If Anne had truly been the brazen, in-your-face floozy, taunting the older woman at every turn, contrasting her youth and beauty to Catherine's dumpiness, why would she have hidden herself away?

At the Christmas celebrations a year later, Anne appeared at the king's side at a banquet, around the time when the papal legate observed Henry "kissing Anne and treating her in public as though she were his wife." After the banquet, she refused to show her face again. The misogynistic view of this behavior is that it was a ruse to tempt him away from Greenwich Palace to Anne's lodgings at York Place, the future Whitehall Palace, where, indeed, he dutifully followed. But was it really scheming manipulation on her part? Or pure discomfort that a man was kissing and fondling her in public with his faithful wife—her beloved former employer—in the next room?

Certainly, Anne must have feared running into the queen and her supporters at every turn. Clearly, she hated being whispered about as the king's selfish, slutty mistress. Couldn't it be that she withdrew from court to spare herself these hideously painful scenes, to maintain a shred of decorum? When Henry proposed to her, Anne had expected an annulment and honorable marriage in a reasonable amount of time. Instead, as the years ticked by and no annulment arrived, she found herself with no marriage, no children, constantly hurting the queen she had served and loved, and a reputation as bad as that of the Whore of Babylon.

In November 1529, according to Chapuys, she railed at Henry, "I have been waiting long and might in the meanwhile have contracted some advantageous marriage, out of which I might have had issue. . . . But alas! Farewell to my time and youth spent to no purpose at all." A year later she "wept and wailed, regretting her lost time and honor, and threatening the King that she would go away and leave him." Anne found herself in an unenviable situation. It is possible—perhaps even likely—that over the course of her six-year engagement to the king she regretted it indeed. But there was no backing out.

Realizing the pope would never annul his marriage, Henry decided to make himself supreme head of the Church of England and marry Anne. Though accounts are conflicting, they probably wed in secret in November 1532, with another ceremony in January 1533. If the earlier date is true, it is likely that the porno-tart-queen Anne Boleyn was a thirty-one-year-old virgin until she married. In June 1533, six months pregnant, she was crowned queen of England.

As queen, Anne focused on helping the poor, promoting religious reform, and sponsoring impoverished scholars. She and her advisors held meetings, drew up plans, and drafted documents that she presented to the king, supporting her proposals with evidence. Her chaplain William Latymer wrote that Anne "favored good learning so much" that she paid for the education of numerous underprivileged youths at Cambridge University. In 1535, the English classical scholar John Cheke wrote that the queen would fund any poor student her chaplains vouched for. She even founded a grammar school, free to those students without the means to pay.

Anne ordered billowing quantities of canvas and flannel from which she and her ladies sewed clothing for the poor. She carried a purse of coins with her to dispense to the needy wherever she went. She sent servants to the towns surrounding the royal palaces to discover those truly suffering poverty and dole out money. While all royals supported—and still do support—worthy causes as a part of their jobs, Anne was personally involved in her philanthropy, working to choose worthy recipients. Unlike the selfish hellcat we know from fiction and movies who loved to prance around in bejeweled gowns and flirt, Anne gave substantial amounts of the income she had as queen to the poor. The legend of Anne Boleyn, however, would have us believe she had no real interest in helping the less fortunate. Her one goal was to outshine Queen Catherine's charity, which had also been substantial.

Admittedly, Anne wasn't all sweetness and light. If we can believe even some of the gossip Chapuys heard, at times she must have had a volcanic temper and a tongue sharp enough to scrape paint off walls, just as her daughter Elizabeth I would have. Frustration fueled her temper. Having waited six years, she was finally queen of England, yet foreign countries refused to acknowledge her, and most Englishmen did so only to prevent the king from killing them. Her first, vaunted pregnancy resulted not in the desired male heir, but only in another girl child, Elizabeth.

It is unclear what role Anne played in Henry's appalling treatment of his first wife and their only child, Mary. The king separated mother and daughter in 1531, sending them to live at different establishments. They would never see each other again. Henry requested that Catherine give up her jewels for Anne's use. He declared his marriage to Catherine null and void, and Mary a bastard. In 1533, Mary's royal household was dissolved, and she was sent to live with her younger half-sister, Princess Elizabeth, where Mary's lowly status as a bastard contrasted with the royal status of the baby.

Did Anne encourage Henry to such cruelty, goading the mild-mannered soul to do things against his will? Or was Henry, furious at being disobeyed as he always showed himself to be, insistent upon his revenge? Chapuys believed Anne was behind every malicious act of Henry's, and many modern historians have eagerly taken up the *blame the woman* theme. "It is she who now rules over, and governs the nation; the King dares not contradict her," Chapuys wrote.

It is hard not to laugh at the image of a shrinking violet Henry VIII meekly accepting Anne's cruel and outrageous orders. Indeed, even after Anne's death, the king subjected his daughter Mary to horrific treatment, viciously grinding her down to the point where the motherless young woman abjectly acknowledged herself a bastard. Even Chapuys couldn't blame this on Anne, what with her moldering in two pieces in an arrow chest. (The king hadn't bothered to order a coffin for her execution. With the queen bloody and mutilated on the scaffold and nowhere to put her, the Tower of London staff had to run around on the world's most bizarre scavenger hunt for a box, any box, *quick*.)

One thing is certain: Anne's reaction to news of Catherine's death refutes much of the misogynistic framework of her story. Her notorious legend would have you believe that in January 1536, when word arrived at court of Catherine's death, she and the king clad themselves in yellow as a sign of jubilation. But if you study Chapuys's letter—the source of this story—you will find that he mentions only the king wearing yellow. If the despised concubine Anne had worn yellow, surely the ambassador would have reported it. The king, and the king alone, wore yellow and was delighted. Chapuys wrote, "After dinner the King entered the room in which the ladies danced, and there did several things like one transported with joy."

It seems that instead of dancing in yellow, Anne retired to her private chapel, locked the door, and cried, according to a French diplomat, Jean de Dinteville. Even Chapuys reported she cried: "Some days ago I was informed from various quarters," he wrote, "which I did not think very good authorities, that notwithstanding the joy shown by the concubine at the news of the good Queen's death, for which she had given a handsome present to the messenger, she had frequently wept." Naturally, Chapuys put a negative spin on the sources and the queen's weeping. If Anne had indeed cried, he wrote, it was due to "fearing that they might do with her as with the good Queen."

Chapuys added that Anne sent her grieving stepdaughter Mary a message "that if she would lay aside her obstinacy and obey her father, [Anne] would be the best friend to her in the world and be like another mother, and would obtain for her anything she could ask, and that if she wished to come to court, she would be exempted from [serving her]." Admittedly, Anne's timing wasn't the best. But she felt compelled to reach out to the bereaved girl and offer an olive branch.

Anne had a miscarriage on the day of Catherine's funeral, perhaps her third. The king, always livid at not getting his way (as if Anne could control the gender of her children and had miscarriages just to spite him), was flirting with other women, which caused her to berate him in loud arguments. Her life was going off the rails.

And Anne had powerful enemies at court. Many disliked the religious reforms Henry had introduced and blamed her. Others deplored a sharp-tongued, loud-mouthed woman, of no royal blood, wielding so much power. Her most potent enemy, however, was chief minister Thomas Cromwell. They had been allies in the king's divorce from Catherine. By 1535, though, they disagreed on several important policies regarding religious reform and foreign alliances. On one occasion, Anne, losing her temper, even threatened Cromwell to his face to have him beheaded. Keenly aware of the king's disappointment in Anne for lack of a son, as well as his interest in one of her ladies-in-waiting, a demure little thing named Jane Seymour, Cromwell concocted a story of Anne's infidelities with five men, including her own brother.

As soon as Cromwell showed him the "evidence" of her adultery, Henry didn't even bother to ask her about it. She was arrested and, within three weeks, beheaded. Chapuys explained that "It was [Cromwell] who . . . had planned and brought about the whole affair." In other words, Henry didn't instruct Cromwell to destroy Anne. Cromwell came up with a plan to remove her as a threat, using Henry's increasing irritation with her as his tool.

Two weeks after her arrest, on May 14, 1536, Cromwell wrote to the English ambassadors in France that the queen had lived such a disgraceful life that her ladies-in-waiting could no longer hide her crimes. Yet Cromwell's report of a dissolute lifestyle squarely contradicted William Latymer's observations as Anne's chaplain. The queen wanted to ensure her ladies led moral lives, he wrote, and would "rebuke" and "sharply punish" those who did not. She instructed all the members of her court to avoid "infamous places . . . evil, lewd, and ungodly brothels" and attend chapel daily. She warned her chaplains not to indulge in "pampered pleasures, nor licentious liberties or trifling idleness, but virtuous demeanor, Godly conversation, somber communication and integrity of life."

Anne has been lucky in one regard; since her own era, few, if anyone, really believed her guilty of adultery. Surrounded by a bevy of ladies day and night, it would have been problematic to have had one lover, let alone five. Many of the dates when, according to the charging documents, Anne supposedly committed adultery were clearly incorrect: according to court records, she was either at another palace on that date or she was recovering from childbirth. None of her ladies were charged as accessories to Anne's crimes, which would have been the case if they had aided and abetted a treasonous adulterer, and, indeed, Henry permitted most of them to serve his next queen. But Henry, sociopathically self-centered, eagerly accepted the accusations about Anne as a quick and easy means to be on to the next wife.

Archbishop Thomas Cranmer, who had worked with Anne on religious reform, wrote, "I never had better opinion in woman than I had in her, which maketh me think, that she should not be culpable." Even her archenemy Chapuys wrote that she was "condemned upon presumption and certain indications without valid proof or confession."

The clues that show another side to Anne have been there all along. They have been mentioned in books about her but never focused on as they prove inconvenient to the misogynistic story of slutty villainess. Let's sum them up. At first, she ran away from the king—for a year—unwilling to be his mistress because she loved Queen Catherine. Once she accepted Henry's proposal of marriage, she couldn't bear to see Catherine. She hated the king fondling her in public. She wept and railed that her reputation was ruined, that she could have been happily and honorably married with children by now. As queen, she helped the poor, provided countless scholarships, and encouraged religious reform. She held a decorous court, where bad behavior was punished. And when Queen Catherine died, Anne locked herself in her room and cried.

Most historians agree that Anne's crime was not adultery but ambition. It is perhaps not unexpected that the self-made men at Henry's court—the butcher's son Thomas Wolsey and the brewer's son Thomas Cromwell, both of whom became the most powerful men in the country after the king—have never been accused of overweening ambition. They have usually been admired for their hard work, intelligence, and cunning to rise so high. Yes, yes, Wolsey was incredibly greedy, with numerous luxurious palaces, and had a mistress even though he was a cardinal. And to do the king's bidding, Cromwell crafted edicts that ended up killing thousands of people. But never mind that. Boys always get a pass. Both men are seen as effective public servants.

When a woman, however, uses intelligence, charm, and hard work to improve her lot, she is accused of the crime of ambition, of scheming and playing the slut to sate her greed and selfishness. Anne Boleyn stood at the fatal crossroads of religious upheaval, political power plays, a sociopathic husband, and misogyny. And a woman daring to rise high in the world, up and out of her assigned place, risks crashing right back down. Without her head.

"A MONSTER OF SELFISH AMBITION"

When thirty-three-year-old Queen Catherine de Medici served as regent of France while her husband King Henri II was off at war in 1552, she was widely praised as "gifted with extraordinary wisdom and prudence," according to the Venetian ambassador. "There is no doubt that she would be very capable of governing." She met with the council, raised money for the army, and negotiated with ambassadors. Soon, the envoy reported, "She is so much loved that it is almost unbelievable."

One of her jobs was to provision the army. "We arranged yesterday another bargain for twenty thousand loaves a day," she wrote her husband. "At the same time, I assure you that everyone who has arrived recently from our camp say they have met a large number of wagons carrying bread, flour and wine."

When a Spanish army slaughtered half the French forces at Saint-Quentin in August 1557, many Parisians fled for the safety of the countryside. Catherine, however, went to Parlement and persuaded them to grant her 300,000 francs and 60,000 men to continue the fight. "She expressed herself with so much eloquence and feeling that she touched all hearts," the Venetian ambassador reported, "and made well-nigh the whole Parlement shed tears of emotion. All over the city nothing was talked about with such satisfaction." From then on, the king relied on Catherine to advise him on foreign affairs.

But after the death of her husband in a 1559 jousting accident, aggressive male relatives took power for themselves, ruling for her sickly fifteen-year-old son King François II as they threatened to send the country reeling into civil war. When the young king died a year later, Catherine deftly positioned herself as regent for her second son, ten-year-old Charles IX, sidestepping the men to their utter and impotent fury. She had not been invited to rule by men; she had, in a shockingly unladylike manner, seized power.

The nobility insisted that a male relative—and all eligible candidates were either treacherous or imbecilic—become regent while Catherine focus on taking care of her children. The closest male relative, a cousin, King Antoine of Navarre, First Prince of the Blood, was a bombastic, fickle moron who changed religion at the drop of a hat for his personal advantage.

But clearly, it would be better to have an incompetent man at the helm than an effective woman, especially a foreigner. While eighty years earlier Anne de Beaujeu had been somewhat palatable as unofficial regent of France, being a French princess, Catherine was Italian. To make her ambitions more acceptable, she initially cloaked them in acceptably submissive female terms. Soon after the death of François, she called a meeting of the council and said, "Since it has pleased God to deprive me of my elder son, I mean to submit to the Divine Will and to assist and serve the King, my second son, in the feeble measure of my experience." She would "keep him beside me and govern the state, as a devoted mother must do."

She was submitting herself to God's will by serving. She was a poor feeble woman, but a devoted mother. A few months later, she proclaimed herself Governor of the Kingdom: "Catherine by the grace of God, Queen of France, Mother of the King." That did not go down as well.

The Venetian ambassador wrote his senate, "This is the present state of France: a very young king without experience or authority; a Council rife with discord; all power residing in the hands of the queen, a wise woman but frightened and irresolute and always merely a woman; the King of Navarre, a very noble and courtly prince but inconstant and with little experience in public affairs; as for the people, they are all divided into factions."

Yet the queen mother outwitted her enemies and stayed in power as her second son—and then the third—proved utterly incompetent to deal with civil war, plagues, floods, bankruptcy, and famine. "Her aim is always to remain in power," the Venetian ambassador wrote, as if that were a bad thing when Catherine's mentally deranged son King Henri III spent all night writing memoranda and the next morning burning them. Or when, ignoring one national crisis after another, he wafted around in elaborate costumes he designed, with baskets of puppies hanging around his neck.

In 1587, when enemy armies surrounded Paris, the Spanish ambassador wrote of Henri III, "The king has done nothing but dance and masquerade during this carnival without cessation. The last night he danced until broad daylight, and after he had heard Mass went to bed until night. He then went to his Capuchin monastery where he is refusing to speak or to see anyone."

With her diplomatic skills, the queen mother was in a unique position to negotiate treaties, tamp down rebellions, and make glittering promises to those who laid down their arms. She sought to arbitrate between Catholics and Huguenots (French Protestants) to avoid all-out religious genocide and scorched-earth civil war. Suffering gout, rheumatism, toothaches, and lung ailments, she rode tirelessly back and forth across France, including, at the age of sixty, an arduous eighteen-month journey to negotiate with Huguenot leaders face-to-face. Sometimes she held meetings from her sick bed. She believed herself to be the only royal advisor who told her sons the truth in no uncertain terms. For instance, on November 25, 1579, she wrote the king, "You are on the eve of a general revolt. Anyone who tells you differently is a liar."

Historians have called Catherine "a monster of selfish ambition," "blinded by her ambition," "insatiable in her ambition," a woman who put "power before affection." Did she enjoy wielding power? Without a doubt. Would she have preferred to have healthy, sane sons who ruled with strength and competence? Most likely. Catherine realized her sons would probably not have been able to stay on the throne without her. Not only did French Catholics and Huguenots wage war against each other, but powerful noble families on both sides brought foreign forces onto French soil.

Early in her widowhood, Catherine wrote to her daughter Elisabeth, queen of Spain, "My principle aim is to have the honor of God before my eyes in all things and to preserve my authority, not for myself, but for the conservation of this kingdom and for the good of all your brothers."

In 1588, Henri III had the leader of the Catholic faction, the popular duc de Guise, stabbed to death in front of him, fomenting a new wave of the civil war Catherine had worked so tirelessly to end. A single action had undone Catherine's decades of negotiation and mediation, of diplomacy and cajoling, of jouncing around France in pain to prevent men from setting the country on fire. It was too much to bear. Days later, as she lay dying at the age of sixty-three, Catherine murmured, "I am crushed to death in the ruins of the house."

With very little reason, as we will see in a later chapter, writers since her own time have blamed Catherine for issuing the command to launch the 1572 Saint Bartholomew's Day Massacre, which ended up killing tens of thousands of law-abiding French Huguenots. Indeed, Catherine's accusers lobbed so many baseless accusations at her for countless crimes—poisoning, pimping, lying, scheming—that she became the Sinister Queen of the Black Legend, also known as Madame Serpent. We can only wonder whether Catherine ran a child sex ring under a popular Paris pizza parlor.

Henri of Navarre, who married Catherine's daughter Marguerite and later became king of France when her line died out, wrote of her, "I ask you, what could a woman do, left by the death of her husband with five little children on her arms, and two families of France who were thinking of grasping the crown . . . ? Was she not compelled to play strange parts to deceive first one and then the other, in order to guard, as she did, her sons, who successively reigned through the wise conduct of that shrewd woman? I am surprised that she never did worse."

"DARK AMBITION AND CRAVING FOR POWER"

Nearly five hundred years after Catherine de Medici, ambitious women are still looked upon with suspicion. Perhaps more than any modern politician, Hillary Clinton has seen unfavorable public reaction to her political goals. Her 67 percent approval rating upon becoming first lady dipped down into the forties when she sought to reform healthcare over the next eighteen months, an effort many believed was a power grab by an ambitious but unelected woman. Her approval ratings soared back to 67 percent in 1998 when she was seen as the long-suffering wife of a philandering husband; finally, she was playing a suitably traditional female role.

When, still in the role of first lady in 2000, Clinton announced her run for a New York Senate seat, her ratings fell to the mid-forties. A native of Arkansas who had bought a house in Chappaqua,

New York, to qualify for the election, she was seen as a carpet-bagger, a person with no ties to the area where they are running for office. Yet even her harshest critics were forced to admit that Clinton was indefatigable throughout her campaign. She visited each of the state's sixty-two counties on a "listening tour," where she met with groups of concerned citizens, with a special focus on Republican voters in Upstate New York. She promised legislation to create hundreds of thousands of jobs, cut taxes for the middle class, expand educational opportunities, improve social security and Medicare benefits, and increase business investments. "I hope New Yorkers will decide it's more important what I'm for than where I'm from," she often said, referring to her carpetbag. Her rising polls got a big boost when her opponent, Republican congressman Rick Lazio, during a televised debate, walked over to her while she was speaking, handed her a soft money fundraising agreement, de-manded she sign it, and wagged a reprimanding, sexist finger in her face as if he were scolding a child. Clinton won with 55 percent of the vote. For seventeen days she was simultaneously first lady and US senator.

In office, Clinton served on five Senate committees, includ-ing the high-powered budget and armed services committees. She worked diligently to help New York's economy recover from the September 11 attacks. As a hardworking senator, she was reelected in 2006 with 67 percent of the vote, winning all but four of the state's counties.

But in 2007, when she announced her run for president, her ratings plummeted back into the forties. Some of the damage may have resulted from the aura of inevitability ringing her candidacy;

according to many pundits, Clinton considered herself the heir to the throne, waiting only for the formalities of a coronation and the anointing with holy chrism. Yet after losing the Democratic nomination to Barack Obama, she graciously accepted his offer of secretary of state, and her approval went up to 69 percent. Once she announced her run for president in 2015, it tanked yet again back into the forties.

In 2017, Clinton told NBC News that "the more a woman is in service to someone else" the more likable she is. At the State Department, she was "in service to my country" and "in service to our president. . . . But when a woman walks into the arena and says, 'I'm going in this for myself,' it really does have a dramatic effect on how people perceive." Her yo-yo ratings reflect public approval when she is "serving" the nation—just as they love women who serve coffee and sandwiches—and how that approval quickly turns to disgust when a woman sets down the coffeepot and runs for executive office.

In her election memoir, Clinton wrote, "I never stopped getting asked, 'Why do you want to be President? Why? But, really—why?' The implication was that there must be something else going on, some dark ambition and craving for power. Nobody psychoanalyzed Marco Rubio, Ted Cruz, or Bernie Sanders about why they ran. It was just accepted as normal."

"HER THIRST BORDERS ON THREATENING"

You might think that a woman vice presidential candidate, running for a supporting role doing whatever the (male) president tells her to do, might meet with some sympathy from the Patriarchy. Not so, alas.

In 1984, Democratic nominee Walter Mondale selected New York congresswoman Geraldine Ferraro as his running mate. Campaigning against the wildly popular incumbent, Ronald Reagan, Mondale made a bold Hail Mary pass, hoping that choosing the first female vice presidential candidate in history would fuel his campaign.

The press secretary of Vice President George H. W. Bush, Peter Teeley, when asked to size Ferraro up as a competitor, said, "She's too bitchy. She's very arrogant. Humility isn't one of her strong points, and I think that comes through." It's strange to think that a woman running for the second-highest job in the land should be humble, perhaps meekly cast her eyes down and shrink into the background, but such was the case. And lest we think that response was just a bit of vestigial sexism back in the day, let's look at what happened in 2020.

Members of Joe Biden's VP vetting committee, including top donors, warned him not to choose Kamala Harris as his running mate, as reported by CNBC. They were upset that she had skewered Biden in the primary debates, criticizing him for opposing the desegregation of public schools by busing in the 1970s. "There was a little girl in California who was part of the second class to integrate her public schools and she was bused to school every day," Harris said. "That little girl was me."

A Chicago-based Biden supporter said, "I think a good number of people closest to Joe are pushing against Kamala, including me.

I don't like her, and I don't like the way she campaigned. She seems not loyal at all and very opportunistic."

According to *Politico*, former senator Chris Dodd of Connecticut, a member of Biden's vice presidential selection committee, was surprised that Harris had no "remorse" for the attack.

It is likely Harris's opponents would have liked her much better had she been deferential, smiling, humble, and pleasant, just thrilled to even be on that stage. They didn't seem to consider that she had been competing for the Democratic presidential nomination as hard as Biden and the other contenders, all of whom had had sharp elbows during the debates. We can only wonder whether Harris's critics would have made the same remarks about her lack of loyalty and friendship if her first name had been Keith instead of Kamala.

In a CNBC report, John Morgan, a Florida businessman and Democratic fundraiser, said, "She would be running for president the day of the inauguration. For me loyalty and friendship should mean something." Other Biden allies agreed, saying that being ambitious, she would want to become president, something—oddly enough—she wasn't supposed to want. That, in other words, the former vice president wanting to become president shouldn't have a vice president who wanted to become president. Fourteen vice presidents had gone on to become president—Biden would become the fifteenth—so it's strange to suggest the person Biden chose as his running mate should have had no aspirations in that direction. (Though, come to think of it, other vice presidents were shy, self-effacing characters, with no presidential ambitions of their own. Just look at Lyndon Johnson and Richard Nixon, the very definition of blushing, bashful fellows who positively wilted under any attention.)

When Mitt Romney chose Wisconsin congressman Paul Ryan as his running mate in 2012, Romney's supporters didn't jump all over Ryan for his political ambitions. On the contrary, the media lauded Ryan as "a young, ambitious beltway insider, with a camera-ready presence." No one suggested Ryan should be more loyal, modest, and deferential, remorseful for anything he ever said that might have sounded critical.

Another ambitious front-runner for Biden's VP pick was Stacey Abrams, a Georgia state politician who narrowly lost the 2018 gubernatorial race. In December 2019, when she was asked by a journalist whether she wanted to be Biden's running mate, she said, "Yes," and then acknowledged how "weird" it was to say so publicly. "I'm a Black woman who's in a conversation about possibly being second in command to the leader of the free world," she explained, "and I will not diminish my ambition or the ambition of any other women of color by saying that's not something I'd be willing to do."

Four months later, she reiterated her position to *Elle* magazine: "Yes. I would be honored. I would be an excellent running mate. I have the capacity to attract voters by motivating typically ignored communities. I have a strong history of executive and management experience in the private, public, and nonprofit sectors. I've spent 25 years in independent study of foreign policy. I am ready to help advance an agenda of restoring America's place in the world. If I am selected, I am prepared and excited to serve."

Naturally, Abrams caught flack for clearly stating her ambitions. Kimberly Ross of the *Washington Examiner* wrote a scathing column calling her "obsessively ambitious. . . . Abrams's desperation for national relevance is her driving force. . . . [H]er thirst borders on threatening. . . . Her feeling of entitlement to the vice presidency just because she's a black woman should annoy individuals on both sides of the aisle."

A Democratic congressman from Missouri, William Lacy Clay, stated he found Abrams's behavior "offensive." "For you to be out there marketing and putting on a PR campaign that way, I think it's inappropriate," he said. When a reporter asked Abrams whether she thought a man would be criticized for expressing his political ambitions, she said, "No."

In contrast to Harris and Abrams, another vice presidential contender, California congresswoman Karen Bass, was praised for her modesty and humility. Bass was described as a nonthreatening "worker bee," according to some media portrayals. One *Politico* article stated: "She's a politician who cringes at having her picture taken and is content to let others grab headlines. . . . In many ways . . . the anti-Kamala Harris." According to the *New York Times*, Biden allies applauded the fact "that she has no interest in seeking the presidency herself." Which, when you come to think of it, was a very strange recommendation, even more so for a president who would be eighty-two at the end of his first term.

The *Washington Post* opined, "Rep. Karen Bass (D-Calif.) is recommended because everyone in the party likes her—sending the message that women who are sweet, cheery and nonconfrontational will be rewarded. These reactions aired in the media perpetuate the notion that only a certain type of woman who does not offend men can be welcomed in the top rungs of power."

HOW TO RUN FOR A JOB YOU ARE NOT SUPPOSED TO WANT

Here's a baffling question: How does a woman pretend to have no ambition when she seeks the most powerful office in the land? These days, there is likely not going to be a sign that God—or the gods—wants her to have it. And unlike a hereditary position, an elected office cannot *fall* on her out of the blue, as it did on Elizabeth Tudor. Today, an ambitious woman has to run for office. Campaign. Debate. Convince people to give her their money and lots of it. Poke holes in the policies of her competitors, most of them men.

Ironically, the only likable female president would be one who doesn't seem to want the job for herself. She must convince voters that—while competent—she is not power-hungry and selfish. Her ambition, if she has such a horrible thing, must be to help people, as any good mother would. At the very least, female candidates would appeal to more voters if they stress that their ambitions are merely to be of service, insist they are just happy to even be considered, talk about all the luck they've had and how many people have helped them along the way.

Though no US female presidential candidate has ever tried it, perhaps her campaign slogan should be: *Don't vote for me. I don't know anything, and I am not worthy to be president.* She might attract a lot of voters who admire her modesty, humility, and reassuring femininity.

This happened to Corazon Aquino, the widow of Benigno ("Ninoy") Aquino Jr., a popular politician who was assassinated in 1983. Aquino, an unassuming housewife and mother, joined the movement to oust then-president Ferdinand Marcos, whom she believed to be responsible for her husband's murder. As calls grew for her to run against Marcos in the next presidential election, she refused to consider it. On a radio program, she said, "Perhaps it would be better not to mention my name anymore." To one reporter, she said, "When Ninoy was alive, I was just a wife and mother. Now I am a widow and mother. There are many, many Filipinos more intelligent than I and who are recognized political leaders." To another, she admitted, "I don't think I'm cut out for it."

When the Filipino management association asked about her economic program, she replied that she was only a housewife with little knowledge of economics. At a luncheon, where she gave a speech titled "My role as a wife, mother, and single parent," she made clear, "I do not seek any political office." The more she protested she was unqualified, not interested, the louder the insistence became for her to run. More than a million people signed a petition begging her to do so for the good of the nation. She finally agreed. When Marcos attacked Aquino for having no political experience, she replied that she indeed had "no experience in cheating, lying to the public, stealing government money, and killing political opponents." Marcos, she said, had too much experience.

In an election rife with fraud and violence, Marcos declared himself the winner, but the armed forces and the Roman Catholic

Church, along with millions of protesting Filipinos, insisted Aquino was the winner. Marcos fled the country. The most unambitious candidate ever had just been elected president. The fact that a woman ran for president must have disgruntled many, yet her modesty, humility, and her traditional role as wife and mother would also have made her more palatable to the Patriarchy. Here was no ball-busting feminist, no loud, strident, ambitious woman, grabbing power for herself. Here was just a sweet widow, hoping to help a suffering nation heal.

And indeed, Aquino's ambitions were reserved solely for her country. After six tumultuous years in office, beset by coup attempts, civil unrest, earthquakes, typhoons, and volcanic eruptions, she declined to run again. Despite her mixed legacy, she had established the Philippines as a democracy with a strong constitution, and often said that no one should be president for life. "I don't like politics," she said. "I was only involved because of my husband."

CHAPTER 3

WHY DOESN'T SHE DO SOMETHING ABOUT HER HAIR?

I get constant comments on the clothes I wear, how fat or thin I am, about my tits, my hair, everything.

—Jess Phillips, British Member of Parliament

As queen of France in an age of breathtaking excess, Marie Antoinette was expected to lead the fashion trends, which the rest of Europe would copy, thereby stimulating the French economy. She wore coiffures three feet high—her hair teased over wooden frames—topped by feathers, ribbons, strands of pearls, and even miniature ships. Her sumptuous wardrobe included hundreds of gowns of silk and satin and jewels the size of hen's eggs. Many of her skirts were so wide it looked as if she had a buffet table under all those yards of shimmering, beribboned silk.

The underground press ridiculed the queen for wearing the

fashions of the day. While excessive female fashions had always been a target of mockery, by the third quarter of the eighteenth century, democracy was in the air like a heady, seductive fragrance. Marie's husband, Louis XVI, had nearly bankrupted the country to help the newly declared United States of America battle France's historical enemy, Great Britain (though many thought it was a stupid idea for a king to help a people rise up against their king; it would give them foolish ideas and *oops!* They were right!). Additionally, France suffered a series of cataclysms—hailstorms, floods, droughts, and famine—which made Marie's outlandish attire look selfish and frivolous while the people starved.

Marie didn't even enjoy the wild fashions she was required to wear. She escaped from them whenever she could, often decamping to a play village on the grounds of Versailles, where she and her ladies dressed simply and milked cows for fun. But this behavior, too, was ridiculed—she was being *unroyal*. In 1783, the portrait painter Élisabeth Vigée LeBrun painted Marie in one of these gowns: white, rather shapeless cotton with a round, ruffled neck, tied by a transparent golden sash at the waist. She wore no jewels, just a wide-brimmed straw hat with some blue plumes. Unwisely, the queen agreed to exhibit the portrait at the salon of the Académie Royale in Paris.

The public was horrified. She looked like a commoner. Where was her royal coiffure? Her stately jewels? To wear such a simple shift—which looked rather like a woman's underwear at the time—was immoral. Immodest. Lacking in the dignity required of a queen. She was breaking down the barriers between the social classes. She was supporting France's traditional enemy, England, a chief supplier of cotton through its colony, India. She was unpatriotic. She would destroy the French silk industry. The selfish woman was pure evil. It is a lesson many modern female politicians have learned: no matter what you do with your appearance, no matter whom you try to please, you will be harshly criticized.

Men, naturally, are rarely chastised for how they look, not in Marie's day, and not now. At Joe Biden's January 2021 inauguration, Bernie Sanders sat stoically in a chill wind, legs crossed, arms crossed, huddled into himself for warmth. He wore a casual brown windbreaker and dark pants. But his knit brown-and-white mittens stood out prominently like cartoon cats' paws. Yes, he was adorable. But honestly, if a woman had worn that windbreaker, those mittens, and sat there like a disgruntled tree toad, critics would have spewed venom. How dare she wear a casual jacket and silly mittens to such an august event? Clearly, she is disrespecting the new president, the American way, democracy, apple pie, baseball, and the Bible. But no. Bernie's attire sparked a meme that went 'round the world, photoshopped into a bazillion photos on social media. Printed on T-shirts, sweatshirts, and stickers, it raised more than $1.8 million in three days for charity.

British prime minister Boris Johnson's hair resembles a bale of straw battered by a tornado; his jackets are wrinkled, his shirts untucked, his collars curled and lopsided, and his ties askew, revealing a short, limp tongue of material beneath. Did his appearance diminish public faith in his ability to steer the nation through Brexit and a global pandemic? Not at all. According to *New York Times* columnist Vanessa Friedman, his fashion faux pas "somehow became badges of credibility that bridged the class gap. He doesn't just break the boring old rules, he blows raspberries at them. His schlubbiness is both a product of his privilege and its antidote. It's a balancing act that leaves his opponents at a loss. And while their physical sloppiness may once have been seen as reflecting a mental sloppiness, in an increasingly airbrushed and filtered world it telegraphs unvarnished truth telling and reality."

"YOU WON'T GET ANYWHERE WEARING SHOES LIKE THAT"

In a 1996 article, "Women MPs and the Media: Representing the Body Politic" in *Parliamentary Affairs*, authors Annabelle Sreberny-Mohammadi and Karen Ross reported that female politicians felt tremendous pressure to be perfectly groomed while their male colleagues benefited from a double standard. "Men colleagues were to be found with lank and dirty hair," the report noted, "dandruff on their collars, stained ties, unsure about the precise positioning of their trouser waistbands (over or under their paunch) and their suits looking as if they had doubled as sleeping bags. If a woman were to appear in a similar state of dishevelment, she would make front-page news that day and questions would be asked about whether she was fit to be a Member of Parliament."

At a 1986 rally in Manila, First Lady of the Philippines Imelda Marcos (she of the thousand pairs of shoes) made a horrifying accusation against Corazon Aquino, her husband's opponent in the upcoming presidential election. "Our opponent does not put on any makeup," Marcos thundered. "She does not have her fingernails manicured!"

Hillary Clinton wrote in her 2017 memoir, "I'm not jealous of my male colleagues often, but I am when it comes to how they can just shower, shave, put on a suit, and be ready to go. The few times I've gone out in public without makeup, it's made the news. So I sigh and keep getting back in that chair, and dream of a future in which women in the public eye don't need to wear makeup if they don't want to and no one cares either way."

After Colorado congresswoman Pat Schroeder (served 1973–1997) gave a speech on policy, voters would come up to her and "want to ask about why you were wearing earrings," she recalled in her memoir, "why you weren't wearing earrings, why do you dye your hair, why don't you dye your hair, why do you wear green? You'd say, 'Can we talk about the speech?'"

Paula Hawks, 2019 chair of the South Dakota Democratic Party, recalled her difficulty in creating an appearance to please the press and her constituents. "You know, should I wear a skirt or a dress or should I not?" she told FiveThirtyEight.com's 2020 survey on issues faced by female politicians. "Will I offend somebody if I wear a skirt or a dress or will I not? Do I need to wear makeup? Do I need to put on more jewelry? Do I need to take off some jewelry? I spent a ridiculous amount of time being focused on my appearance."

Mallory Hagan, a 2018 candidate for the US House of Representatives, told FiveThirtyEight.com, "I recall at one point being at a meeting and talking about myself, my platform and just sharing my candidacy with a group of people. And a woman came up to me afterward saying, 'You won't get anywhere wearing shoes like that.'"

In her long public-service career before serving as prime minister of Australia from 2010 to 2013, Julia Gillard knew that "what you are wearing will draw disproportionate attention. . . . Pleading, 'I like to wear suits' or 'I have been on the road for days' simply did not cut it." As she recalled in her autobiography, "Undoubtedly a male leader who does not meet a certain standard will be marked down. But that standard is such an obvious one: of regular weight, a well-tailored suit, neat hair, television-friendly glasses, trimmed eyebrows. Being the first female prime minister, I had to navigate what that standard was for a woman. If I had appeared day after day in a business suit, with a white camisole and blue scarf, the reaction would have been frenzied—and, I suspect, vicious. Of course, other female politicians have had to work through these issues, too, but none with the spotlight as white hot as it was on me."

She noted that, "Records needed to be kept on what I had worn when. You could not wear something too frequently or, heaven forbid, wear the same thing two years in a row to the same event."

British MP Jess Phillips wrote in her 2017 memoir, "I wish I could say from the dizzy heights of the career ladder that the way women look doesn't matter. I wish I could say that women feeling anxious and judged by their appearance is a teenage phenomenon. It isn't." She recalled a conversation with another female MP who lamented she didn't have the "right sort of body to be on TV." Which had Phillips wondering if Boris Johnson ever got on the phone to David Cameron to complain that the camera made him look plump.

"When Theresa May wakes up in the morning on the cotton-sheeted bed in 10 Downing Street," Phillips wrote, "the first thing she has to think about isn't Russia bombing Aleppo or the fact that the UK currency is spiraling out of control thanks to Brexit; no, she thinks, what am I going to wear in order to face these challenges and avoid comments on my appearance?" Tony Blair, Phillips guessed, would have had no such concerns.

"Every day of our lives, women are told we have to look a certain way," Phillips continued, "our bodies need to be a certain size and shape, and yet if we live up to those standards we have become a massive distraction for the men around us. Theresa May should be able to wear frumpy clothes—a tracksuit if she wants—as long as she can rock up and be decisive, controlled and intelligent."

In an interview with British *Vogue*, Phillips explained that the endless commentary on her appearance was a source of continual frustration. "I get constant comments on the clothes I wear, how fat or thin I am, about my tits, my hair, everything."

Women leaders end up losing valuable time on matters related to their appearance. Julia Gillard wrote, "As Prime Minister, time is not for wasting. Make-up sessions were combined with briefings—usually from media staff. In my world, it was routine to have a press secretary, policy adviser and my chief of staff all yelling a briefing at me over the sound of a hair dryer being used to give my already dry hair a final touch up. I became adept at lip-reading; whoever needed to get across the most information had to be directly in front of me so I could see their mouths forming the words."

Hillary Clinton told the authors of the book *Women and Leadership*, "In the presidential election, if you conservatively say I spent an hour a day for hair and make-up, that's an hour that a male candidate didn't have to spend, and it added up to twenty-four days. It's absurd! Twenty-four days out of my campaign were spent getting ready to go campaign. A man gets in the shower, shakes his head, puts on his suit, which is pretty much the same as everybody wears, and gets out of the door. So, it does breed a certain amount of resentment where you are, like, 'Wait a minute, what am I doing?' It is time-consuming and exhausting."

"DEEP DOWN INSIDE A LOT OF THEM WANTED TO BE FASHION REPORTERS"

The media has been at the forefront for stirring debate on the appearance of female politicians. True, women have had a much wider selection in terms of hairstyles, cosmetics, jewelry, colors, and clothing styles ever since the 1790s, when London fashion icon Beau Brummell tossed aside pink satin knee breeches for black jackets, long pants, white shirts, and cravats, starting a fashion revolution that has changed little between then and now. There is, admittedly, not much to be said about whether a man chooses a blue or red tie. But reams can be written—and are written—about a woman's choice of skirt or slacks, heels or flats, prints or stripes, makeup, hair, and jewelry.

When Julia Gillard became Australia's new deputy Labor leader in 2006, the *Daily Telegraph* reported, "On what should have been one of the proudest days of Gillard's political career, she bungled it with a less than flattering haircut and a frumpy '80s tapestry print jacket. . . . Get yourself a stylist your own age."

Media obsession with Gillard's appearance only got worse

when she became prime minister in 2010. She recalled in her 2014 memoir, "It is galling to me that when I first met NATO's leader, predominantly to discuss our strategy for the war in Afghanistan, where our troops were fighting and dying, it was reported in the following terms: 'The Prime Minister, Julia Gillard, has made her first appearance on the international stage, meeting the head of NATO, Anders Rasmussen, in Brussels. Dressed in a white, short jacket and dark trousers, she arrived at the security organization's headquarters just after 9 am European time and was ushered in by Mr. Rasmussen, the former Danish Prime Minister and NATO Secretary General.'" This article was written by a female journalist, who made no mention of Mr. Rasmussen's suit.

One journalist—Australian-born feminist author Germaine Greer, shockingly enough—commented on Gillard's "big arse." On a TV show called *Q&A*, Greer jeered, "What I want her to do is get rid of those bloody jackets. Every time she turns around you've got that strange horizontal crease, which means they're cut too narrow in the hips. You've got a big arse, Julia. Just get on with it."

Gillard recalled the television footage in which a telephoto lens focused on her rear end as she entered a car, "not something done to male prime ministers. Even before Germaine Greer's attention-seeking outburst about my body shape and clothing, apparently my arse was newsworthy."

Michelle Bachelet, president of Chile from 2006 to 2010 and again from 2014 to 2018, recalled in the book *Women and Leadership* that the only time a South American women's magazine reported on her as president, the story said "something like 'Unbelievable!

In the same week she wore the same suit twice!' I was surprised. They could have written about powerful women, but they chose to write about this. But I also knew that if I changed my clothing too much, I would be dismissed as frivolous."

In 2015, the *Daily Mirror* summed up the changed appearance of Nicola Sturgeon, first minister of Scotland and leader of the Scottish National Party: "She's lost shed-loads of weight. She's got a sleeker, less carroty new hairdo. She's got a natty new wardrobe of suits with matching stilettos and confidence way beyond her abilities." The *Daily Mail*, which called her a "wee woman," published numerous articles solely on her appearance, such as this in 2015: "SNP leader Nicola Sturgeon has left her boxy jackets and severe suits in the past—and she proved her new style credentials with a stunning appearance yesterday morning. The 44-year-old looked particularly glamorous on her way to BBC's Andrew Marr Show in a fuchsia column dress that flattered her slimmed-down physique."

Christine Todd Whitman, the first female governor of New Jersey who served from 1994 to 2001, described to FiveThirtyEight .com what she saw as the innermost longings of political reporters. "It seems like really deep down inside a lot of them wanted to be fashion reporters," she said.

It is likely that many journalists are merely trying to paint a picture for the reader by discussing a female politician's apparel, while there is admittedly not much to say about a man's. Perhaps reporters are looking for a message in the fashion, some symbolism to be teased out of the color of a dress or the cut of a jacket. Whatever the case, media mentions of a female politician's appearance have disastrous effects. Research has shown that any mention at all—even a positive one—diminishes her in the eyes of the public.

In 2013, Name It. Change It., a joint project of the Women's Media Center and She Should Run, an organization dedicated to helping women explore running for public office, released the results of a survey of 1,500 US voters nationwide, along with a sample of one hundred young women, aged eighteen to thirty-five, in which they looked at a hypothetical congressional race between Jane Smith and Dan Jones. Survey respondents read a profile of each candidate, along with sample news stories covering their positions on an education bill. Nothing was said about Dan Jones's appearance; the articles about Jane Smith included either a positive, negative, or neutral description of her appearance, or no description at all.

Naturally, the negative description resulted in a lower rating: "Smith unfortunately sported a heavy layer of foundation and powder that had settled into her forehead lines, creating an unflattering look for an otherwise pretty woman, along with her famous fake, tacky nails."

But the neutral description also resulted in a negative impression among respondents: congressional candidate "Jane Smith dressed in a brown blouse, black skirt and modest pumps with a short heel . . ."

Surprisingly, even a flattering portrait of a female candidate lowered her in the voters' esteem. The description of Jane Smith as "fit and attractive and looks even younger than her age," even though it sounded complimentary, hurt the voters' perceptions of the politician for being in touch, likable, confident, effective, and qualified. Maybe they thought she spent too much time at the gym and salon, instead of taking care of her husband and children. Maybe she wouldn't take care of voters either. She was probably selfish. Ambitious.

Such was the experience of Irene Sáez, the mayor of a major Venezuelan municipality and winner of the 1981 Miss Universe pageant. In 1998, Sáez had a reputation for honesty and efficiency when she ran for president of the country. Tall and attractive with long, wavy golden hair, she commanded a wide lead in the polls until the press trivialized her by commenting on her beauty, clothing, makeup, and hair, shifting coverage away from her ideas and policies, while focusing on the ideas and policies of her male competitors. Frustrated with the press coverage, Sáez took to wearing her hair in a bun—which made headlines. She lost the election, winning a humiliating 3 percent of the vote.

Interestingly, in the Name It. Change It. study, candidate Jane Smith scored highest with participants when there was no mention of her appearance at all. Those who intend a gallant compliment to a woman in public office have no idea that they are, in fact, doing harm. For instance, in April 2013, then-President Barack Obama attended a Democratic National Committee event in Atherton, California, where he introduced various politicians and mentioned their stellar qualifications. When he came to California Attorney General Kamala Harris's accomplishments, he added, "She happens to be by far the best-looking attorney general in the country," with the quite unintended results of diminishing her in the minds of the audience.

THE PANTSUITIFICATION OF POWERFUL WOMEN

A powerful woman's wardrobe choices are fraught with pitfalls. Dark colors can be viewed as sinister and threatening; they make her look bossy and humorless. Pastel colors, on the other hand, may make her look frivolous, wishy-washy, and weak. Bright colors, however, scream for attention, something no woman should do, and show that she is not serious enough. Baggy clothing shows that she is frumpy, unpolished, incompetent, and insecure. But well-tailored clothing might make her look uncaring and cold, and it probably cost too much.

In recent decades, female politicians have Beau Brummelled themselves, choosing the pantsuit, a feminized version of a masculine uniform, as their template. It exudes a sense of no-nonsense power. It's also a way of saying, *Don't look at me. Listen to me.*

In her election memoir, Hillary Clinton explained her sartorial infatuation with pantsuits: "When I ran for Senate in 2000 and President in 2008, I basically had a uniform: a simple pantsuit, often black, with a colorful shell underneath. I did this because I like pantsuits. They make me feel professional and ready to go. . . . A uniform was also an anti-distraction technique: since there wasn't much to say or report on what I wore, maybe people would focus on what I was saying instead."

But sometimes even a prudent anti-distraction uniform can cause controversy. When Julia Gillard went to the province of Queensland in 2011 to see for herself the destruction caused by floods and a cyclone, she decided to wear a dark suit for her joint press conference with Queensland premier Anna Bligh. A plain, dark suit would be fitting for such a somber occasion. And Gillard had seen Bligh wearing a black suit the day before. But when Gillard arrived at the press conference, Bligh was wearing a white shirt, casual pants, and boots. Her bangs flopping into her face, she looked a bit rumpled, as if she hadn't gotten a wink of sleep, which she probably hadn't. Gillard was hammered for looking too polished and serene in the face of such horror.

The press reported, "Yesterday as the floodwaters threatened her state capital, Bligh fronted the media in a utilitarian white shirt, hair looking like she had been working all night. . . . Beside her, Ms. Gillard stood perfectly coiffed in a dark suit, nodding. For women politicians, it is always a fine balance between showing emotion and being perceived as too emotional. Gillard has perhaps erred toward being too cool." Professor Ross Fitzgerald, a Queensland historian and emeritus professor of history and politics at Griffith University, said, "In contrast to Anna Bligh, Prime Minister Gillard has seemed wooden and not caring. I am not saying that she doesn't care; it's just she doesn't appear to care." Commentary on her inappropriate dark suit went on for six weeks.

The question here is: Why didn't Julia Gillard know she should have worn wrinkled clothes, slept in them perhaps, messed up her hair, and smeared her mascara to show how much she cared about the victims? How could she not have known that? And what did this egregious error in judgment say about her ability to run the country? Was this appalling wardrobe choice a sign that she was the wrong person for the job?

Gillard mused about the public reaction if a male prime minister had appeared looking polished and professional in a dark suit at the press conference. "I doubt wearing a suit would, in and of itself, have become an issue, and been equated with his not caring," she wrote in her memoir.

Even if sedately swathed in a pantsuit uniform, women have to deal with necklines, something men do not. In 2007, when Hillary Clinton was discussing the enormous costs of university education, she shocked the world by indicating she had breasts. (Who knew?) "There was cleavage on display Wednesday afternoon on C-SPAN2," wrote Robin Givhan in the *Washington Post*. "It belonged to Sen. Hillary Clinton. . . . There wasn't an unseemly amount of cleavage showing, but there it was. Undeniable. To display cleavage in a setting that does not involve cocktails and hors d'oeuvres is a provocation." The cleavage was indeed seemly enough, a V-shaped black top beneath a pink jacket. Perhaps the provocation wasn't the cleavage so much as an individual with cleavage running for president.

Nor is such neckline journalism limited to the US. In 2008, German chancellor Angela Merkel wore a low-cut black dress to the inauguration of Norway's new national opera house in Oslo. Even though that particular event did involve cocktails and hors d'oeuvres, her choice was enough of a provocation to make the front pages of papers in Germany and across Europe. The UK's *Daily Mail* ran photos beneath the headline "Merkel's Weapons of Mass Distraction." Online, the headline was "Deutschland boober alles." The Iron Frau didn't care. As the unflappable Merkel patiently explained to a German TV station, "It's simply due to the fact that, in Germany, a woman is chancellor."

In 2007, when UK Home Secretary Jacqui Smith spoke of

terrorist bombing attempts in London, the *Sun* reported, "Jacqui Smith, the new home secretary, made her first Commons statement yesterday to the great admiration of some, not so much for what she said as for the amount of cleavage she had on display." The *Daily Mail* described the offending bosom as a "rather middle-aged, squeezed together line of amplitude, about three inches long. . . . A little desperate, if anything, and designed to draw the eye down from the face and slim the chins away. Certainly not sexy." The *Sun* ran an article titled "Best of Breastminster," which rank the "ministerial boobs." The resulting uproar became known as the "tempest in a D-cup."

In 2016, when then British Home Secretary Theresa May wore a stylish red suit to an event, the headlines focused on her "busty budget" and "boob-boosting push-up bra." One journalist mused whether she had been wearing a leopard-print bra to match her favorite shoes.

Female politicians are criticized not just for what they wear but also for how much it cost. When Hillary Clinton campaigned in a Giorgio Armani jacket in 2016, the media noted it retailed for $12,495, and people raged about the cost. By virtue of wearing such an expensive garment, *she was out of touch with regular people.* Perhaps the price tag was not *so* over the top when we consider Clinton was bringing in up to $325,000 per speech at the time. In 2008, the press raked Republican vice presidential candidate Sarah Palin over the coals when it was revealed the Republican National Committee paid $150,000 for her campaign wardrobe, though Palin didn't shop for the clothing herself and had no choice in the matter. In contrast, Donald Trump has never been condemned for his Brioni suits, which run as high as $17,000 apiece.

And let us picture, for a moment, Hillary Clinton on the presidential campaign trail wearing a $150 suit from Macy's. *She's cheap*, the Internet would have roared. *No taste. No self-respect. Unpresidential.*

Noted author and professor of communication Kathleen Hall Jamieson explained that women in power face a double bind, which means *You're damned if you do and damned if you don't.* "Women who are considered feminine will be judged incompetent, and women who are competent, unfeminine," she wrote. "Women who succeed in politics and public life will be scrutinized under a different lens from that applied to successful men."

Even successful men who stray wildly from the accepted uniform of dark suit and tie receive their fair share of criticism, however. In 2018, on a trip to India, Canadian prime minister Justin Trudeau decided to wear traditional dress to meet male Bollywood icons, who he assumed would be wearing traditional dress. In a modern version of an O. Henry story, the actors, however, wore plain dark blazers, the traditional dress of Western politicians. There was the Canadian prime minster, shining like the sun in a golden brocade *sherwani*, a long coat, looking for all the world as if he was wearing a very expensive Halloween costume. The media attacked him for "playing dress-up." "Too Indian even for an Indian," mused one Indian magazine.

President Barack Obama, known for his refined good taste in clothing, told *Vanity Fair* in 2012, "You'll see I wear only gray or blue suits. . . . I'm trying to pare down decisions. I don't want to make decisions about what I'm eating or wearing. Because I have too many other decisions to make." He made an exception, however. On August 28, 2014, he wore a tan suit to a press conference regarding US plans to respond to ISIS in Syria.

Conservative commentators roundly pilloried him for it. Republican congressman Peter King of New York said of Obama's crime, "There's no way, I don't think, any of us can excuse what the president did yesterday." Borrowing from the title of Obama's 2004 Democratic National Convention address and his 2006 book, *The Audacity of Hope*, critics called his fashion faux pas "the audacity of taupe."

THE HARROWING HASSLE OF HAIR

These days, the media can be cruel to men and women alike. Donald Trump's hair has been compared to "an obedient Persian cat holding very still on the top of his head," "the male equivalent of a push-up bra," and a "decomposing ear of corn." Mostly, though, when the hair of powerful men hovers anywhere near the brink of normalcy, no comment is made, and when it is, it is often done almost endearingly. (*Why, just look at Boris Johnson's unique bale of hay hair, so cute, clever, and creative!*) Women, however, are another matter.

"I certainly got a lot of critiques from the media and constituents about my hair," said Claudine Schneider, who in 1980 was the first woman elected to Congress from Rhode Island. "That was extremely disconcerting, to say the least," she told FiveThirtyEight .com. "'Why don't you do something with your hair, or why don't you cut your hair, or why don't you curl your hair?' Everyone had a different idea of what kind of hairstyle I should have."

Female candidates have developed a version of pantsuits for their hair: helmet hair. If her hair is too long or curly or feminine, she is seen as not serious, flighty, vain. If it is too short and spiky, she is seen as tough, hard, uncaring, maybe even a lesbian, which, in certain conservative corners, is still regarded as sexually depraved and causes concern among more left-leaning voters about a candidate's ability to win. A smooth bob, somewhere between earlobe and shoulder, usually attracts the least mention, helping the press and public focus on the candidate's message.

Jacinda Ardern, who became prime minister of New Zealand in 2017, recalled how she agonized over the burning question of how to style her long hair in a televised election debate. "I remember thinking, what can I do so my appearance is not the subject of commentary?" she said in an interview for the book *Women and Leadership*. "I decided to wear my hair up because then it wouldn't get in my face, it wouldn't distract." Unfortunately, the fact that her long hair didn't distract became a distraction. She recalled, "After the debate, all these messages started coming into my office about how much people disliked the fact that I had worn my hair up, and it became a real point of contention."

Even worse, in May 2020, a New Zealand television host, Ryan Bridge, asked Ardern if she dyed her hair. She told him it was not a polite question. Shoving his foot further down his throat, he replied, "Looks good, Prime Minister. And I only mention the gray hair because you are the prime minister, and it does tend to age people. No harm intended, alright?"

Bridge was highly offended by the shower of criticism he received. "I cannot emphasize the number of messages that personally I received from haters around the globe defending Jacinda Ardern, like I'd somehow insulted all womanhood by asking about her gray hairs," he said, clearly feeling victimized by his innocent, well-meaning remarks.

Hillary Clinton has been criticized for just about every hairstyle she has worn. Her headbands in the early nineties got a universal thumbs-down. A *Forbes* columnist deemed the butterfly clip she wore in 2012 as secretary of state too girlish and lacking in "gravitas." (We are not quite sure what *gravitas* means, but it seems to have something to do with testicles.)

As secretary of state, Clinton bounced ceaselessly around the world, often flying overnight and landing in steamy weather. She took to pulling her hair back in a scrunchie, a cloth-covered elastic band. As one of her aides told *Elle* magazine, "As a chick, it's a big pain in the butt. The weather is different, and you're in and out of the plane. [The staff] gets off that plane looking like garbage most days, but she has to look camera ready. She said the reason she grew her hair long was that it's easier. She has an option."

In 2012, when Clinton spoke about a gender equality and women's empowerment initiative in Cambodia, the press reported that her staff wanted to ban her scrunchies. Clinton joked that she should have called her book about her years as secretary of state, "The Scrunchie Chronicles: 112 Countries and It's Still All About the Hair." She told CNN she didn't care much about the reaction to her appearance. "If I want to wear my glasses, I'm wearing my glasses. If I want to wear my hair back, I'm pulling my hair back. You know at some point it's just not something that deserves a lot of time and attention."

In 2015, Matt Drudge of the Drudge Report claimed that Clinton was wearing a wig (evidently a symbol of her lack of authenticity) at various events and posted several pictures of her with no scalp showing. Clearly, earth-shattering news.

In a 2013 *New York Times* column, Maureen Dowd wrote of Clinton, "She has ditched the skinned-back bun that gave her the air of a K.G.B. villainess in a Bond movie. . . ." Patricia Marx of *The New Yorker* wrote, "Hillary Clinton changed her hairstyle one million times, and the one way she didn't try was the one way that works."

Then again, Clinton found that the fascination with her appearance had its uses when she was first lady. "If we ever want to get Bosnia off the front page," she quipped in 1995, "all I have to do is change my hair."

Compare the coverage of Clinton's hair to that of Senator Bernie Sanders. During his presidential campaigns, the Vermont Democrat often appeared at the podium in a rumpled suit, his hair like a demented Chia Pet. Great was the speculation as to whether he was in possession of a comb.

In a 2015 *Al Jazeera* article, several experts opined on the glories of Sanders's bedhead look. "It makes him authentic, a rather important attribute in this presidential cycle when voters are attracted to unscripted candidates who act and sound real," said Bruce

Newman, a marketing professor at DePaul University, who authored a book on how candidates relay their message in the twenty-first century.

"I think it's refreshing," said Symone Sanders, national press secretary for Sanders's 2016 presidential campaign. "It's just Bernie going out there and being Bernie. . . . He doesn't think his hair is pertinent news to Americans." Well, just tell that to Hillary Clinton.

"It makes perfect sense for him not to comb his hair," said Lara Brown, professor of political management at George Washington University. "His personal style, which tends towards the disheveled, reflects his anti-materialist, egalitarian beliefs."

Fair enough. But would a female presidential candidate with crazy disheveled hair be called "authentic" and "anti-materialist"? Or just crazy and disheveled? Wouldn't Bernadette Sanders be seen as an unkempt, scolding grandmother who didn't care enough to run a comb through her hair and was, therefore, undeserving of the highest office in the land? Picture how people would handle Hillary Clinton with a frizzy white perm, or Kamala Harris with waist-length Rastafarian dreadlocks. Surely they would be called "unpresidential." (What does that even mean? Perhaps that the individual in question is in possession of two breasts and a uterus?)

In October 2020, the *Washington Times* castigated Representative Alexandria Ocasio-Cortez (D-NY) for visiting a Washington hair salon with prices higher than the writer thought she should have been paying with her own money (though the prices were, alas, fairly standard in DC). The headline read: "EXCLUSIVE: Self-declared socialist AOC splurges on high-dollar hairdo." The salon charged $80 for a haircut, the article crowed, and $180 for lowlights. Contrast this Gucci socialist with former Attorney General Jeff Sessions, a modest and humble soul who got his hair cut for $20 by the Senate barber.

Women of all political parties, once they stopped laughing, must have wanted to say, *Duh, of course men get their hair cut almost for free while women have to take out a mortgage.* Feminist writer Jessica Valenti tweeted in response to the story, "Sorry you don't get to create beauty standards that require women to spends hundreds or thousands a year to be considered presentable and then hate us for it." And let us imagine, for a moment, that AOC had gone to a Hair Cuttery equivalent. Wouldn't the *Washington Times* have railed about her stinginess? Criticized a less-than-top-of-the-line hairstyle?

In September 2020, Speaker of the House Nancy Pelosi was blasted for seeing her hair stylist privately at his salon in spite of COVID-19 restrictions. Yet we can only imagine the uproar if she appeared at the speaker's podium with white roots and limp locks.

Few have noticed that President Joe Biden, at the age of seventy-nine, has more hair on top of his head than he did thirty-five years ago. Back in 1987, one reporter did notice the change and asked the then senator about it. Biden replied, smiling, "Guess I've got to keep some mystery in my life." And that was the end of that.

There is a sliver of space in which a female candidate's appearance is not judged harshly: when she is winningly attractive but not a sex siren. Her hair, clothing, makeup, and jewelry are respectable but not worthy of much comment. Kamala Harris has the good fortune and genetics to fit into that tiny space perfectly. Her makeup is understated; her jewelry, often pearls, classic. She usually pairs dark, plain pantsuits with lighter blouses, with not the slightest indication that she is in possession of a pair of breasts.

There have been rare occurrences when the less-than-telegenic appearances of powerful women have not been excoriated. Perhaps it's no coincidence that the two who come to the top of mind held political offices outside of the US. Prime Minister Golda Meir of Israel had her start in politics in the 1940s and 1950s, raising hundreds of millions of dollars to set the new nation on its feet and welcome Holocaust survivors from across Europe. As prime minister from 1969 to 1974, she was uncompromising, fiercely defending Israel's interests. Focused solely on the country's survival against great odds, she had absolutely no concern about her looks. While today's female politician sporting a gray bun, no makeup, dumpy dresses, and orthopedic shoes would draw harsh criticism, especially in the US, Meir was seen as a tough-talking but lovable Jewish grandmother, a beloved character.

Ellen Johnson Sirleaf was a plump sixty-seven-year-old grandmother when she became president of Liberia in 2006. At the hands of previous administrations, she had survived imprisonment and several close calls with execution. Sirleaf wanted it known that she was tough enough to take on the male-dominated culture of corruption, to put warlords in their place and stem violence (it was estimated that some 70 percent of Liberian women had been raped). A finance expert with a Harvard graduate degree and an impressive employment history at the world's top banks, she also wanted it known she would efficiently tackle the nation's severe economic problems. As a mother of four and grandmother of ten, she was a revered figure in traditional West African dress in a culture that cherishes the wisdom and experience of older matriarchs. Perhaps because of this tradition, the press did not criticize her for her age, weight, sexual history, voice, ambition, or clothing.

"AN ABILITY TO BE OURSELVES"

There are times when female politicians use their clothing to make a powerful statement; they *want* it to be written about in the press and noticed by the public. Such was the case at the 2018 State of the Union address, when almost all the Democratic women in Congress wore black in solidarity with the #MeToo movement. At the 2017, 2019, and 2020 addresses, they wore white, the color of the suffragettes, stating loud and clear that they stood together in the fight for women's rights under a president who had bragged about assaulting women. Viewers—and the president himself, standing at the podium—couldn't miss the sea of women in white.

When the Democrats regained control of Congress in 2019, and Nancy Pelosi was sworn in again as Speaker of the House, the *New York Times* commented on the color of her dress—hot pink—resulting in a torrent of angry tweets from women who thought the remark disempowering. But the *Times*'s fashion writer Vanessa Friedman begged to differ. "I don't think there's any question," she wrote, "Ms. Pelosi picked a hot pink dress for her swearing-in both because she knew it would make her stand out in what was still a room full of dark suits, and because of the symbolic nature of the occasion: a color traditionally associated with delicate femininity had become a color associated with a seat of power. That's a strategic and savvy choice, and to take notice of it is to acknowledge the multidimensional chess game Ms. Pelosi is playing, not to demean her."

Pelosi is not alone in using a pink dress to trumpet forth a message. When Judge Amy Coney Barrett appeared at her Senate confirmation hearing for her nomination to the Supreme Court on October 12, she wore an attractive dress in magenta. Perhaps she was sending a message highlighting her femininity (in her charismatic religious group People of Praise she is a "handmaid," after all, whatever that is, and I'm not sure we really want to know). Maybe she was signaling her independence, stating that she would not meekly fit the mold supporters and opponents alike had prepared for her. Maybe her message was—now here's a novel thought—that she just liked the dress.

Whatever the case, no sooner had DC lawyer Leslie McAdoo Gordon seen the offending garment than she harrumphed to her 25,000 Twitter followers, "Women lawyers & judges wear suits, including dresses with jackets, for work. It is not a great look that ACB consistently does not. No male judge would be dressed in less than correct courtroom attire. It's inappropriately casual." Twitter erupted in an uproar as Gordon was accused of sexism, defended herself, and was accused of sexism again for defending herself. Barrett must have been delighted at her swearing-in, not only because she was on the Supreme Court, but even more importantly, from that moment on she would be expected to wear a black robe to work.

When Margaret Thatcher rose to power in the UK political scene in the 1960s, executives at the PR firm Saatchi & Saatchi helped her craft a new image. She chose power suits, often in bright shades of blue. But two pieces of advice she fiercely resisted: to ditch her trademark pearls and her handbags. Her advisors feared these feminine accessories would make her look weak, incompetent, even female. But Thatcher didn't care. She used them to proclaim her strength. Yes, she had a uterus. Yes, she loved clothes. Anyone who didn't like it could go straight to hell. Her handbags, in fact, entered the British lexicon as a symbol of strength. "To handbag" someone means to argue loudly and finally get them to do what you want them to, as Thatcher did in cabinet meetings when she banged her bag on the table.

The second woman to become British prime minister, Theresa May, was somewhat nonconformist in her wardrobe choices, wearing bold colors, above-the-knee dresses, leather pants, and kitten heels. In 2002, as chairwoman of the Conservative Party, she gave a controversial speech to her fellow Tories warning them that they were becoming the "nasty party" for petty party infighting and demonizing minorities. The next day, the *Daily Telegraph* devoted one-third of its front page to a picture of Theresa's stylish leopard-print shoes, accompanied by the headline "A Stiletto in the Tories' Heart."

The *Daily Mirror* saw May as embodying a "dominatrix fantasy, with her formidable, finger-wagging, headmistress act. . . . The sight of Theresa May in kitten-heeled leopard skin 'don't f*** with me' shoes was enough, apparently, to bring tears to the eyes of red-blooded Tories on the first day of the party conference."

The day after Prime Minister David Cameron announced May would take over his position in July 2016, the British newspaper the

Sun ran a front page taken up mainly with a photo of Theresa May's feet, encased in wild leopard-skin pumps—with red leopard skin in between the heel and the toe—and rhinestone buckles. "Heel, Boys," the headline read, over small photos of male politicians' heads, with the subtitle "New PM Theresa can reunite Tories & deliver Brexit." Another article had the headline: "Theresa shows her steel: Metal toe-capped shoe-wearing PM stamps her authority on the Tories."

In October 2017, the BBC was criticized for focusing on the prime minister's heels during a Brexit report. One viewer who complained said that the video "sends all the wrong messages, deflects from what she was saying, shows no respect for her position, and reinforces gender stereotypes."

May has shrugged off media commentary of her clothing and, indeed, has often brought up the subject of her footwear in interviews. "I like clothes, I like shoes," she told the crowd at the 2015 Women in the World summit. "One of the challenges for women in politics, in business, is an ability to be ourselves. You can be clever and like clothes, you can have a career and like clothes."

Sebastian Payne wrote in the *Financial Times*, "The prime minister has a strict trio of personal topics that she is prepared to discuss in public: shoes, clothes and cooking. Beyond that, she is content to let voters judge her on achievements."

May, who is somewhat introverted and reserved, has often used fashion—in particular, her vast array of shoes—as an icebreaker. It makes her relatable, approachable to the public. And if she discussed fashion with a journalist, they might not ask her about Brexit or why she never had children, always a sore subject. Her shoes were so unusual—often in leopard skin, contrasting colors of lizard skin, adorned with buttons and bows—it got to the point where, whenever the prime minister entered a room, all heads would look down at her feet.

During the 2020 campaign and before, Kamala Harris could often be seen at public events or traveling wearing Converse Chuck Taylor sneakers, sending the message she was a woman on the go, working for the American people. In 2020, she told the *Cut*, "I run through airports in my Converse sneakers. I have a whole collection of Chuck Taylors: a black leather pair, a white pair, I have the kind that don't lace, the kind that do lace, the kind I wear in the hot weather, the kind I wear in the cold weather, and the platform kind for when I'm wearing a pantsuit."

Fashion symbolism was front and center at the 2021 inauguration of Joe Biden. The women in Biden's entourage—Kamala Harris and the Biden family members—wore a rainbow of bright, hopeful colors. First Lady Dr. Jill Biden wore a shade of peaceful ocean blue. A representative for the designer, fashion label Markarian, explained that the shade of blue worn by the first lady was chosen to "signify trust, confidence, and stability."

Vice President Kamala Harris wore purple, a tribute to Shirley Chisholm, the first Black woman to run for president under a major party, who used the color in her campaign. Former First Lady Michelle Obama appeared in a bold shade of raspberry. First granddaughter Natalie Biden wore a bubblegum-pink coat and scarf. Youth Poet Laureate Amanda Gorman glowed in a coat the color of butter. The variety of bright colors symbolized the beauty of American diversity.

And when Joe Biden gave his first address to a joint session of Congress on April 28, 2021, for the first time ever, two women sat behind the president as vice president and Speaker of the House. Eschewing the dark suit uniform, Harris wore a blush-colored suit with a golden camisole; Pelosi appeared in pale blue. Perhaps the message in their clothing was: "We're women. We're here. Get over it."

"DON'T YOU DARE STEP INTO THE PUBLIC SPHERE"

First ladies, having no political power of their own, are usually ignored by the Misogynist's Handbook, especially if they observe the patriarchal rules by giving elegant tea parties, supporting their husbands, and promoting a worthy cause. Jacqueline Kennedy redecorated the White House. Lady Bird Johnson beautified America's highways by planting wildflowers. Nancy Reagan just said no to drugs. Laura Bush pushed for childhood literacy. Such first ladies have been applauded by the press and the people. Those who stepped outside of these acceptable bounds have been vitiated.

Eleanor Roosevelt was savaged for her buck teeth, her weak chin, her freakish height, her sensible shoes, her flowered hats, her large handbags, her fluty voice, and several other areas in which she, clearly, fell short of popular expectations. Many other first ladies were not classically beautiful, either—they were middle-aged or older, some of them plump—but were not abused in the press for it. The media and many Americans condemned Eleanor not because of her appearance but because of the power she wielded. She had stepped out of her assigned place as supportive spouse and gracious hostess hoisting a teapot. She was changing the world, working hard to lift up the impoverished and oppressed, and it pissed people off. A national columnist called her "impudent."

Shortly after the 1933 inauguration, she toured Washington's "alley slums," where people lived in tenements without running water, and launched a campaign for decent housing across the country. Her husband, Franklin, consulted her on all his appointments and major legislation. She wrote a newspaper column six days a week for twenty-seven years and appeared regularly on radio programs advocating for her causes.

But . . . those *teeth*.

As her biographer Blanche Wiesen Cook said, "There are those who focus on her teeth and voice and other cartoon characteristics, long before they reveal how much they despise her politics, most notably her interest in civil rights, and racial justice, or in civil liberties and world peace."

Roosevelt shrugged off the criticism. "Every woman in public life needs to develop skin as tough as rhinoceros hide," she said. (Rhinoceros hide is two inches thick, a kind of armored plating.)

As first lady, Hillary Clinton, too, raised public ire for her nontraditional role. In January 1993, President Bill Clinton named her to chair a task force on National Health Care Reform. Two years later, she helped create the Office on Violence Against Women at the Department of Justice. At the 1995 Fourth World Conference on Women in Beijing, Clinton spoke against the abuse of women in countries around the world. "It is no longer acceptable to discuss women's rights as separate from human rights," she stated. In 1997, she helped pass the State Children's Health Insurance Program and initiated the Adoption and Safe Families Act. In 1999, she helped pass the Foster Care Independence Act, which doubled federal funds to assist teenagers aging out of the foster care program. During her tenure as first lady, Clinton traveled to seventy-nine countries to assist with US diplomatic efforts.

But . . . her *hair*.

Terry O'Neill, president of the National Organization for Women from 2009 to 2017, said the appearance of both male and female politicians is judged, but women are critiqued far more harshly than their male colleagues. Referring to an unflattering Drudge Report article on Hillary Clinton's appearance, in May 2012, O'Neill told *USA Today* that it was "really saying to all women, don't you dare step into the public sphere, we will savage you for what you look like."

CHAPTER 4

THE DANGERS OF FEMALE HORMONES

Oh! Menstruating woman, thou'rt a fiend
From which all nature should be closely screened.

—Anonymous

"Yฺou could see there was blood coming out of her eyes, blood coming out of her—wherever," Donald Trump said, referring to Fox News commentator Megyn Kelly, who had grilled him during a 2015 Republican debate for calling women fat pigs. It was a clear reference to menstruation, an insinuation that Kelly's raging hormones had turned her into an angry, irrational woman who attacked poor harmless Donald Trump without reason.

Similarly, as soon as Trump heard that Kamala Harris was Joe Biden's VP pick in August 2020, he tweeted that she was "a mad woman," "extraordinarily nasty," and "so angry." A Trump campaign fundraising email called her "the meanest" senator. Here, clearly, was a woman who could not control her emotions, a woman sabotaged by her own estrogen.

Even the liberal-leaning press has bought into the ancient theory of estrogen contamination. In 2016, *Time* magazine published a well-meaning piece suggesting that post-menopausal Hillary Clinton was the perfect age to become president, intimating that she had put all the horrors of menstruation and menopause behind her. Otherwise, we can infer, she might have been a screeching harpy, ready to use the nuclear codes to destroy the planet in a fit of hormone-induced fury.

Though Hillary Clinton was probably unperturbed by full moons or hot flashes when she ran for president, in 2022 many of our twenty-four female senators and 121 female representatives are in their thirties, forties, and fifties. To our knowledge, they do not morph into monsters every twenty-eight days or try to set the Capitol on fire in a menopausal rage. No one has suggested penning them in a big red tent once a month in the Patriarchal Statuary Hall, where several dozen scowling white marble men on pedestals could glare down on them.

The myths surrounding menstruation—and its evil twin, menopause—have plagued women for thousands of years and show little sign of letting up. A woman's menstrual cycle was a monthly reminder of her power to bring forth life, a raw, primeval power men did not have, and could never have, no matter how brave or skilled in battle, no matter how many dangerous animals they speared and brought back as meat. And this mysterious female power must have scared the bejesus out of them. For instance, the Babylonian Talmud stated that "if a menstruating woman passes between two [men], if it is at the beginning of her period, she will kill one of them."

Some of our ancient forefathers believed that menstruating women could control storms and lightning. The natural philosopher Pliny the Elder (23–79 CE) wrote in his book *Natural History*, "There is no limit to the marvelous powers attributed to females. For, in the first place, hailstorms, they say, whirlwinds and lightning even, will be scared away by a woman uncovering her body while her monthly courses are upon her. . . ."

Even *hailstorms* were scared of her and got the hell out of there.

Pliny added—apparently with a straight face—"If a woman strips herself naked while she is menstruating, and walks round a field of wheat, the caterpillars, worms, beetles, and other vermin, will fall from off the ears of corn." In an era before chemical pesticides, menstruating women would have proven very useful indeed in an agricultural society. An additional benefit: "Bees will forsake their hives if touched by a menstruous woman." She could then collect the honey sting-free. But there was a downside. If she touched a cauldron, he wrote, the linen boiling in it would turn black, and if she so much as put a finger on a razor, its edge would become blunted.

One method to reduce fear is to deride, belittle, and shame that which is feared, which is what ancient cultures did by transforming women's awe-inspiring kinship with nature into a mental aberration. "Hysteria" is derived from the Greek word for *uterus*. And according to the ancient Greeks, if a woman's period was late, her uterus would travel around her body—a phenomenon known as wandering womb—and she would become hysterical (as anyone would, what with their uterus climbing up between their kidneys and aiming straight for the heart). Roman law limited women's legal rights due to the "the imbecility, the instability of the sex" from all those mood swings.

In Christianity, mood swings and imbecility devolved into something unbearably disgusting. Saint Jerome (ca. 342–420 CE) wrote: "Nothing is so unclean as a woman in her periods; what she touches she causes to become unclean."

In the thirteenth-century *Rule of Anchoresses*, Christian women were commanded to despise the "uncleanness" of their own bodies. "Art thou not formed of foul slime?" the rule thundered. "Art thou not always full of uncleanness?"

Penitential regulations laid down in the seventh century by Theodore, archbishop of Canterbury, forbade menstruating women to take Communion or even enter a church as they were believed to "pollute" the altar of God. As late as 1684, certain churches instructed women in their "fluxes" to keep their polluting slime outside the church door.

Let us take a moment here to give earnest consideration to the subject of bodily waste. It's safe to say that nothing that ended up in the bottom of a chamber pot—feces, urine, and menstrual blood—smelled like roses, but why was menstrual blood singled out for such vicious loathing? Why were (male) people full of revolting feces and urine allowed to go to church and pollute the altar of God but (female) people with menstrual blood were considered foul and gross and kept outside?

Pope Innocent III (reigned 1198–1216) confidently declared

that menstrual blood was "so detestable and impure that, from contact therewith, fruits and grains are blighted, bushes dry up, grasses
die, trees lose their fruits, and if dogs chance to eat of it, they go
mad." Which totally explains the origin of rabies.

In 1298, the Synod of Würzburg commanded men not to approach a menstruating woman. (Though we wonder how they
would *know*. Did women wear red armbands? A scarlet *M* pinned
on their dress? Some questions, alas, are forever lost to history.)

In the Middle Ages, many believed that menstrual blood would
burn up any penis it came in contact with. A child conceived during
menstruation would be either the devil's spawn, crippled, or, even
worse, red-haired from all that blood (though it's difficult to understand how a child could be conceived at all if the penis had been
burnt up).

In the sixteenth century, medical authorities continued to
believe that "demons were produced from menstrual flux." The
French royal physician André du Laurens worked himself up to
a pitch of horror and asked: "How can this divine animal, full of
reason and judgment, which we call man, be attracted by these obscene parts of woman, defiled with juices and located shamefully at
the lowest part of the trunk?" (Perhaps the good doctor would have
been less scandalized if the vagina had been located in a more respectable location, somewhat higher up. In the middle of the forehead, perhaps?)

Imagine, then, the disasters that would befall a kingdom if one
of these maniacal, foaming-at-the-mouth creatures, defiled with
loathsome juices and periodically leaking a substance so toxic it

shriveled penises and birthed demons, became its ruler. Not merely as regent for a warring husband or young son—which was a temporary position and could be taken away if she went completely berserk—but as the crowned, anointed sovereign. To prevent such a catastrophe, the Frankish king Clovis, in the year 500 CE, compiled a civil law code known as the Salic Law, the best-known tenet of which excluded women from inheriting the throne. Over time, Salic Law was adopted by several other countries in Western and Central Europe.

Ironically, Salic Law, created to prevent menstruous, monstrous female rulers from dragging the country into unnecessary wars, caused wars that would otherwise have been unnecessary. When Charles VI of Austria died in 1740, his heir was his twenty-three-year-old daughter, Maria Theresa. For eight years, various male relatives aligned themselves with European powers to fight for the throne, resulting in hundreds of thousands of casualties. "It would not be easy," she wrote, "to find in history an example of a crowned head acceding to government in more unfavorable circumstances than I did myself."

After all the bloodshed, Maria Theresa retained her crown, ruling wisely for forty years. She instituted educational, medical, and financial reforms, transforming her kingdom into a modern state and raising its standing from a Central European backwater to that of a great and respected power. The empress herself was admired throughout Europe as "the glory of her sex and the model of kings."

England had no Salic Law, though many Englishmen must have heartily wished it did. The first female "ruler" was the twelfth-century empress Matilda—the widow of the Holy Roman Emperor and sole legitimate heir of English king Henry I. She actually never ruled the entire country, nor was she crowned. Her time was balefully known as "the anarchy," a period of civil war as her forces battled those of her cousin, Stephen of Blois, for the throne. The impasse was resolved when Matilda retired to her domains in France, and her son Henry Plantagenet became Stephen's heir.

Four centuries passed before the royal line again ran out of men. Henry VIII's daughter Mary I ruled for a turbulent five years, during which time she burned hundreds of heretics and married the despised Philip II of Spain, who dragged England into his war with France.

Upon Mary's death in 1558, England faced the terrifying dilemma of yet another female sovereign with the accession of Elizabeth Tudor to the throne. And so, as soon as Elizabeth's coronation celebrations died down, there must have been some apprehension as to what would happen with a woman at the helm. Everything depended on whom she would marry.

Marry she must, of course. For one thing, if a healthy young woman did not have sex, it was believed that "the unruly motions of tickling lust" could cause "naughty vapors" to rise from her private parts to her brain, resulting in pimples and insanity. Elizabeth's secretary of state, William Cecil, believed that sex in the blessed state of matrimony would reduce the queen's "dolours and infirmities as all physicians do usually impute to womankind for lack of marriage." A husband would keep her as sane as it was possible for a woman to be, particularly during those times of the month, and guide her erratic policymaking.

But, as we all know, naughty vapors notwithstanding, Elizabeth's solitary forty-five-year reign was the most peaceful and successful, both domestically and internationally, in the history of the nation. It was so successful, in fact, that it scared men. So they wrote her off as an exception, a miracle, a one-off, an example of God's grace. And sank even further into their derision of women's hormones.

Even the enlightened monarch Frederick the Great of Prussia derided female rule as blighted by hormonally induced melodrama. "A woman is always a woman," he wrote in 1774, "and, in feminine government, the cunt has more influence than a firm policy guided by straight reason."

A century later, little had changed. For six months in 1878, the *British Medical Journal* ran a series of articles debating whether a menstruating woman could turn a ham rancid by touching it. In 1919, the Hungarian-American pediatrician Béla Schick (1877–1967) believed that menstruating women released substances called "menotoxins," a kind of airborne poison gas exuding through their skin, which killed plants nearby. As late as 1974, in the respected, peer-reviewed British medical journal *The Lancet*, one researcher claimed that a permanent wave would not "take" to a woman's hair during menstruation.

While some women suffer from cramps and hot flashes more than others, it has proven impossible to shake the age-old beliefs that female hormonal fluctuations transform women into special-needs individuals unfit for positions of responsibility. For instance, in 2014, the British National Union of Teachers, in a bizarre bid to be helpful, decided that menopausal teachers had special needs due to their "memory loss," "not being able to finish a sentence," and "feelings of anxiety." These teachers required additional breaks and improved sanitation.

It was certainly refreshing in 2015 when a study published by Imperial College London examined the destruction caused not be estrogen but by testosterone. The research found that excess testosterone can make men act irresponsibly and impulsively (we are *shocked*), a possible cause of the 2008–2009 banking crisis in an industry dominated by men. Researchers found high testosterone levels induced risk-taking, impulsivity, and irrationality. (Perhaps we should call them "testotoxins"?) And, as we know, a heady supply of testosterone is not limited to a few days a month.

"RAGING HORMONAL IMBALANCES"

Men have been using menstruation and menopause to justify keeping women out of government for a very long time. When American women campaigned for the vote in the late nineteenth and early twentieth centuries, men often responded that women were too emotional to play a role in government. In 1873, Orestes Brownson, a prominent American intellectual and writer, argued, "As an independent existence, free to follow her own fancies and vague longings . . . without masculine direction or control, [a woman] is out of her element, and a social anomaly, sometimes a hideous monster."

In the late 1890s, suffragist Anna Howard Shaw attended the Democratic National Convention in Baltimore. Having been told for years that women were unfit to vote because of their uncontrollable emotions, she was shocked to see men "screaming, yelling, and shouting at their people," she reported. "I saw men jump upon the seats and throw their hats in the air and shout, 'What's the matter with Champ Clark?' Then, when those hats came down, other men would kick them back into the air, shouting at the top of their voices: 'He's all right!!!' Then I heard others screaming and shouting and yelling for 'Underwood!! Underwood, first last and all the time!!!'"

Naturally, the men didn't consider themselves to be hysterical or prey to sinister lunar influences. Shaw wrote they saw their impassioned behavior as "patriotic loyalty, spending manly devotion to principle." They jumped up and down in their seats, yelled and screamed, threw their hats, and ran in circles until five o'clock in the morning. "I have been to a lot of women's conventions in my day," Shaw observed, "but I never saw a woman knock another woman's bonnet off her head as she shouted: 'She's all right!'"

By 1970, when American women had had the vote for fifty years, Congresswoman Patsy Mink of Hawaii believed that the day was soon coming when a woman could become president of the US. When Mink expressed her view to Dr. Edgar Berman, a member of the Democratic Party's Committee on National Priorities, he replied in no uncertain terms that women suffered "raging hormonal imbalances" that rendered them unfit to lead. "Suppose that we had a menopausal woman President who had to make the decision of the Bay of Pigs or the Russian contretemps with Cuba at the time?" he asked, referring to dangerous international crises JFK handled in the early sixties. She could be "subject to the curious mental aberrations of that age group." In other words, women having their periods or going through menopause might escalate a war unnecessarily—as Lyndon Johnson did—or start one from scratch, as George W. Bush did.

Berman barreled blithely onward, "Now, anything can happen, knowing women, psychologically during this period, or during their lunar problem. Anything can happen from going up and eating the paint off the chairs. . . ." At this point, Congresswoman Mink stopped him. Soon after, she got him fired. Undeterred, Berman wrote a book called *The Compleat Chauvinist: A Survival Guide for the Bedeviled Male*.

Minutes after the 2020 vice presidential debate, Florida Republican senator Marco Rubio tweeted a GIF of ten missiles exploding from the ground and zooming skyward with the advice, "Think hard about what you just saw . . . then decide who you want just one heartbeat away from the Presidency." He seemed to be indicating that Kamala Harris—indeed, women in general—could easily become hormonally unhinged when dealing with foreign affairs and national security, conveniently forgetting the fact that we have had three female secretaries of state. In addition, according to a 2019 article on the Council on Foreign Relations blog, "Women's Participation in Peace Processes," "The participation of civil society groups, including women's organizations, makes a peace agreement 64 percent less likely to fail."

A 2016 article on the Council for Foreign Relations website called "Women, Peace, and Security" stated that "when women's parliamentary representation increases by five percent, a country is almost five times less likely to respond to an international crisis with violence. Within countries, women's parliamentary representation is associated with a decreased risk of civil war and lower levels of state-perpetuated human rights abuses, such as disappearances, killings, political imprisonment, and torture."

The article adds, "Women often take a collaborative approach to peacemaking and organize across cultural and sectarian divides. . . . Including women at the peace table can also increase the likelihood of reaching an agreement because women are often viewed as honest brokers by negotiating parties."

Despite such evidence, millennia-old beliefs about goggle-eyed, hormone-crazed women ready to blow up the world are still with us today. In a 2009 survey, government professor Jennifer Lawless at American University reported that 25 percent of Americans believed that "most men are better suited emotionally for politics than are most women."

"It's not surprising, because labeling women 'emotional' has long been a tried-and-true tactic to undermine women," Amanda Hunter, executive director of the Barbara Lee Family Foundation, which advances women's representation in American politics, told *U.S. News.* "Men have been doing it—and some women as well—since the women's suffrage movement," when opponents said women were too hormonally unreliable to be given the vote.

"HER EYES ONCE WATERED ON CAMERA"

Because of the age-old belief of women's fluctuating hormones and wild moods, those in public office are often judged far more harshly for expressing emotion than men doing the exact same thing. Take crying, for instance. When male politicians weep, they are perceived positively as authentic, vulnerable, and in touch with their feelings. When women do so, they are seen as demented train wrecks.

John Boehner, the US Speaker of the House from 2011 to 2015, was so notorious for crying in public that he was often called the "Weeper of the House." His watery blue eyes would gush like opened fire hydrants at the least provocation. Irish music made him weep. He blubbered his way through singing "America the Beautiful." While taping a *60 Minutes* segment touring the Capitol, he recalled how he mopped floors at his father's bar in the 1950s and started to sob. Small children reminded him of his eleven brothers and sisters, causing his eyes to redden and flood with nostalgic tears. He admitted to crying while watching the beauty of a sunset.

In 2015, when Pope Francis gave an address from the Capitol balcony, speaking of God's love, Boehner—a devout Catholic who had served as an altar boy—stood beside him making funny faces, wiping his eyes and nose, and repeatedly licking his lips. Yes, yes, the media made fun of him. But no one questioned his mental stability or suggested he was a hormonal mess unsuited for his job. Perhaps to make this point, when outgoing Speaker Nancy Pelosi turned the podium over to him, she left a large box of Kleenex on it. As she handed Boehner the gavel, he kissed her on her cheek and cried.

During the 2016 election, Hillary Clinton was campaigning in a New Hampshire coffee shop when a woman asked her a question and she replied, "I just don't want to see us fall backward as a nation," as her eyes filled with tears. "This is very personal for me. Not just political. It's not just public. I see what's happening. We have to reverse it." The clip was run and rerun on television, as news commentators dissected and analyzed if she truly was crying or if she was being manipulative by pretending to cry to gain sympathy. Commentators debated whether a momentary loss of emotional control would help Clinton in the polls by displaying human warmth and dispelling her robotic reputation, or harm her by showing a woman cracking under the strain, untrustworthy with the nuclear codes.

As she wrote in her election memoir, "My eyes glistened for a moment and my voice quavered for about one sentence. That was it. It was the biggest news story in America. It will no doubt merit a line in my obit one day. 'Her eyes once watered on camera.'"

In 1987, when Congresswoman Pat Schroeder of Colorado announced she would be discontinuing her presidential run, she cried for a few seconds at the podium. "Suddenly I thought I had let everybody down," she explained to *Ms.* magazine five months later, "and they didn't let me down. I felt so crappy." She quickly pulled herself together and finished her speech.

The next morning, newspapers all over the country ran front-page pictures of Schroeder crying. Political analysts debated whether her tears doomed any future that women might have in politics. Her opponents observed that such a show of emotion was alarming in a presidential candidate. A *New York Times* columnist tsk-tsked, "She's the stereotype of women as weepy wimps who don't belong in the business of serious affairs." Another wailed, "What a devastating indictment of this girl's character." (Note: the girl was forty-seven years old.)

Schroeder was irritated by the unflattering coverage. "You don't see anybody saying never again can a man be governor of

New Hampshire because John Sununu cried so hard he couldn't even finish his speech when he was saying good-bye," she told the *Los Angeles Times*. "Or never again could a man run for president because I think every single one of them has shed tears in public now. And then people say, 'We don't want somebody's finger on the button that cries.' Okay, you could debate that. I don't want anybody's finger on the button that doesn't cry. But everybody that I've known whose finger has been on the button has been publicly crying."

In her memoir, Schroeder pointed out that President Lyndon Johnson sobbed through a White House civil rights ceremony. President Ronald Reagan cried when he left office. President George H. W. Bush was known to cry at sad movies. In his farewell speech after President Jimmy Carter fired him, Secretary of Health, Education, and Welfare Joseph Califano Jr. wept openly. When George Washington gave a farewell dinner to his Revolutionary War generals, he and General Knox "suffused in tears . . . embraced each other in silence." At the end of the dinner, there wasn't a dry eye in the house. "Now crying is almost a ritual that male politicians must do to prove they are compassionate," Schroeder wrote, "but women are supposed to wear iron britches."

On May 24, 2019, Prime Minister Theresa May spoke for nearly seven minutes to announce her resignation. When she reached the last four words—"the country I love"—her voice quavered with emotion. Perhaps she really was about to burst into tears, because she slammed her notebook shut with a kind of angry defiance and ran inside Number 10 Downing Street. But she had certainly not cried at the podium.

Naturally, headlines trumpeted the fact that the prime minister had cried. For example, the *Times* led with "Theresa May in tears as she resigns" and the *Sun* with "Prime Minister's teary farewell statement." Those who had not watched the video would have assumed May had been standing there weeping and slobbering.

Six weeks later, May told the *Telegraph*, "If a male Prime Minister's voice had broken up, it would have been said 'What great patriotism, they really loved their country.' But if a female Prime Minister does it, it is 'Why is she crying?'"

Helen Clark, who became the second female prime minister of New Zealand in 1999, recalled a male politician breaking down in tears as he spoke publicly about his child's struggle with drug addiction. And the media treated him with compassion. "They can do it, and they are seen as human," Clark told the *Whig-Standard* in 2018. "If women do it, we're weak." She described a time when her political opponents verbally attacked her during a political meeting. "And I had tears rolling down my cheeks," she said. "They still play that bit of film. They still play it. And play it pejoratively."

Congresswoman Pramila Jayapal of Washington State told the *Cut*, "I like to say that it's a good thing when we cry because policymaking is better when you have emotion about it. I think this whole myth that you have to be dispassionate, that you can't feel things, was constructed by men in power and is an excuse for why we have bad policies. But when you feel the pain of a family not having health care or losing their home or being in poverty or losing a child to police violence, you are more inclined to address it."

The Misogynist's Handbook has been so successful in hard-wiring us with sexism that we are often blind to the double standard at work. After Kamala Harris's professional, level-headed questioning of Attorney General Jeff Sessions during a Senate hearing, Trump advisor Jason Miller said, "I think she was hysterical. She was trying to shout down Attorney General Sessions, and I thought it was way out of bounds." (She was not shouting.)

The same critics—including Jason Miller himself—rarely seemed to mind Donald Trump's daily bouts of foaming-at-the-mouth ranting and raving truth-free melodrama. As Twitter star The Volatile Mermaid tweeted in 2018, "Imagine if a woman president got on Twitter every morning to complain about people being mean and unfair to her. Weak. Hysterical. Shrill. Bitch. Unfit to Lead."

CHAPTER 5

THE ALARMING SHRILLNESS OF HER VOICE

It took me quite a long time to develop a voice, and now that I have it, I am not going to be silent.

—Madeleine Albright, first woman
US Secretary of State

The silencing of woman is deeply rooted in history. In one of the earliest Western poems, Homer's *Odyssey*, written in the early seventh century BCE, Odysseus's teenaged son rebukes his mother, Penelope, for speaking in her own house in front of her own visiting suitors. "Mother, go back up into your quarters and take up your own work, the loom and the distaff. . . . Speech will be the business of men," he commands. And, wordlessly, she does.

Mary Beard, professor of classics at University of Cambridge, has traced the general revulsion to women's voices across the centuries in her engrossing book *Women & Power*. Ancient Greek and Roman literature, she pointed out, emphasized the harmony and beauty of the male voice in contrast to the shrill tones of the female. In societies that valued feats of arms above all other qualities, a deep voice symbolized bravery, a high-pitched one cowardice.

As the Greek writer Plutarch argued in the early second century CE, "A woman should as modestly guard against exposing her voice to outsiders as she would guard against stripping off her clothes. For in her voice as she is blabbering away can be read her emotions." Which are also terrifying, of course. Saint Paul, generally accepted to be the founder of Christianity, vigorously insisted that women shut up. In 1 Timothy 2:12, he writes, "I do not permit a woman to teach or have authority over a man; she must be silent."

For thousands of years, women's voices have been perceived as irritating in some way, usually because they are considered "shrill." In the Book of Isaiah in the King James Version of the Bible, the name of Lilith, an ancient Middle Eastern female spirit of chaos, is translated as "screech owl," perhaps as much for her shrill voice as for her similarity to a bird of prey. In Greek mythology, the god Zeus turned a princess named Io into a cow to hide her from his jealous wife, Hera, but one added advantage must have been that Io could now only moo. When she ran into her father on the lane, Io had to scrape her name in the dirt with her hoof to make herself known—luckily, her name was really short.

The water nymph Daphne, fleeing from her suitor Apollo, called on her father to help her resist the god's rape. Her father, a river god, quickly turned her into a tree, whereupon she could only helplessly wave her branches to attract attention, making no sound other than a gentle rustling of her leaves. Apollo loved her more than ever. According to *Bulfinch's Mythology*, he said, "Since you cannot be my wife, you shall assuredly be my tree. I will wear you for my crown. I will decorate with you my harp and my quiver . . . and, as eternal youth is mine, you also shall be always green, and your leaf know no decay." And Daphne, now changed into a (completely mute) laurel tree, "bowed her head in grateful acknowledgment."

In the sixteenth and seventeenth centuries, women whose speech was deemed "riotous" or "troublesome" would be sentenced to wear for a period of hours a scold's bridle—a kind of iron chastity belt for the mouth. A metal cage surrounded the lower face, and a bridle bit was pressed against the tongue, preventing speech.

Sadly for the American novelist Henry James (1843–1916), the scold's bridle was out of fashion by his time. As Mary Beard noted, in his essay "The Speech and Manners of American Women," he lamented the ruination of society caused by women's voices. American women, James wrote, were destroying human speech, causing it to devolve into a "generalized mumble or jumble, a tongueless slobber or snarl or whine of the emancipated women," that sounds like the "moo of a cow, the bray of the ass, and the bark of the dog." (The sexist tactic of comparing women to animals—something less than human—is explored in Chapter 9.) He ruminated darkly about women's "twangs, whiffles, snuffles, whines, and whinnies," which sounds eerily similar to the "shrill, caterwauling, shrieking, yowling and screeching" that *Time* magazine journalist Jay Newton-Small described in 2016 as "all associated with women—not men."

Queens, at least, didn't have to put up with shrillness critiques. Centuries ago, a queen's voice—even the voice of a monarch as powerful as Elizabeth I—was rarely heard by anyone other than her courtiers and servants. The select few chosen as local representatives might have heard her address Parliament every few years, and she may have given a speech on rare occasions from a balcony where without microphones few in the crowd below would have heard her. Given the aura of sanctity associated with God's anointed royalty, those who did must have felt as blessed as if they had heard the voice of an angel from heaven above.

A queen consort wasn't supposed to speak publicly at all. As Anne de Beaujeu, princess of France, wrote in her 1498 book, *Lessons for my Daughter*, other than bearing children, the primary duty of a queen was silence. If she uttered a word, it should be to comfort and serve those in need, never to entertain or debate. (This was a rule that, while running the country for her younger brother, Anne certainly never applied to herself.)

Frederick the Great felt that even reigning queens should shut the hell up. Whenever he heard that a female ruler such as Catherine the Great or Empress Maria Theresa had uttered insults against him, he instructed his minister to give a sermon on the text of Saint Paul, "Let the woman learn in silence."

Today's politicians have no choice but to speak publicly and often. Their campaigns comprise television, radio, and the Internet. If a candidate's voice is not heard loud and clear around the country, there is no chance she will get elected. And as soon as it is heard, criticism erupts over its quality.

In the 1970s, novelist Norman Mailer said that Bronx-born feminist congresswoman Bella Abzug had a voice that "could boil the fat off a taxicab driver's neck."

During the 2008 US presidential campaign, TV commentator Glenn Beck devoted an entire segment of his show to the horrors of Hillary Clinton's voice, which, he said, was "like an ice pick in the ear." It's "nagging," makes you "envy the deaf," and "makes angels cry." He added, "She could be saying, 'All right, Glenn, I want to give Glenn Beck $1 million,' and all I'd hear is, 'Take out the garbage.'" *InfoWars*'s Alex Jones played a video comparing the former first lady's laugh to that of a hyena. Even liberal media took their potshots. Bob Woodward on *Morning Joe* suggested she "get off this screaming stuff." Joe Scarborough, who had been interviewing him, agreed. "Has nobody told her the microphone works?"

"I, and many other people, do find Hillary Clinton's voice to be shrill. In fact, it sounds like a cat being dragged across a blackboard a lot of the time," said one guest on Fox, during a debate in March 2016 about whether there was sexism in the "shrillness" complaints leveled against Clinton.

On September 30, 2007, no less than three *New York Times* articles ripped into her laugh, one of them calling it "the Clinton cackle." The next day, ABC, CNN, Fox News, and MSNBC ran segments on it. Commentators failed to mention that each word bellowed out of Bernie Sanders's mouth is like a pot with a Brooklyn accent banging on your head, or that Donald Trump's monotone shouting a torrent of words is like a vicious assault by random Ping-Pong balls. When Bernie and Donald yell, and thunder, and chastise, and point and make fists, and gesticulate wildly, their au-

diences praise them for their political passion and compelling oratorical style.

But when a Hillary Clinton or an Elizabeth Warren comes even
close to it, men think, "Why is my mother mad at me?" or "She
reminds me of my angry third-grade teacher." A forceful woman
mirrors a Jungian archetype—an ancient, universal symbol buried
deep in the human subconscious—which strikes horror and fear
into the hearts of men. She's a nagging wife demanding to know
where they've been. A Freudian nightmare of a pissed-off mother
insisting they clean their room. A scolding schoolteacher nun ready
to whack them with a ruler. A scowling librarian striding to their table with her finger to her lips. Nurse Ratched coming to force-feed
them their pills. Subconsciously, they believe they are in trouble,
and it outrages them. Their hackles rise. Any woman who causes
this reaction, which offends all patriarchal norms of men reigning
supreme, must be aggressive, angry, obnoxious, bossy, and complaining. A shrew, a fury, a harpy, a termagant.

Perhaps because of extreme Jungian archetypal sensitivity,
many women speaking in a normal tone of voice are perceived as
shouting. During the 2016 presidential campaign, Donald Trump
accused Hillary Clinton of "shouting" about women's issues. "Well,
I haven't quite recovered," Trump said on *Morning Joe*. "It's early in
the morning—from her shouting that message."

Bernie Sanders, too, indicated Clinton was shouting when she
spoke forcefully about gun control on the campaign trail in 2015.
"All the shouting in the world" would not keep "guns out of the
wrong hands," he said. Clinton took the remark to be sexist, yet
another criticism of women's voices. "I haven't been shouting," she

said at a 2016 presidential debate. "But sometimes when a woman speaks out, people think it's shouting."

Amanda Hunter of the Barbara Lee Family Foundation told the *Washington Post*, "Women are often accused of shouting. You're hard-pressed to find a male politician being criticized for his voice volume."

As a result of the unending criticism of their voices, many women in the public eye are naturally concerned about how their voices are perceived. Such was the case for Chilean president Michelle Bachelet, whose political rise was particularly moving given the fact that in 1974 her father had been put in prison by Chile's former dictator, General Augusto Pinochet, where he was tortured and died. "The day I became the Minister for Defense," she said in an interview for the book *Women and Leadership*, "everybody thought I was thinking about my father and the historical chain of events. But you know what I was actually thinking of? I was thinking I couldn't speak like a girl—I couldn't have this young, feminine voice, I was concerned about having a strong voice from the beginning."

Many female politicians have taken professional voice training to allay the criticism. In the 1960s, Margaret Thatcher, an up-and-coming British politician, realizing her voice was shrill, decided to seek professional help. Sir Laurence Olivier arranged for his voice coach at London's National Theatre to help her lower her pitch. Of course, even that didn't save her from criticism. As prime minister twenty years later with a low, well-modulated voice and excellent diction, she was told she sounded "too headmistressy," as if she were lecturing a naughty child for sticking a wad of chewing gum to the bottom of a schoolroom desk. Prime Minister Thatcher was unrepentant. "I have known some very good headmistresses who have launched their pupils into wonderful careers," she said, shaking off the criticism as ducks do water.

Hillary Clinton, too, decided to get professional voice training. "After hearing repeatedly that some people didn't like my voice," she wrote in her election memoir, "I enlisted the help of a linguistic expert. He said I needed to focus on my deep breathing and try to keep something happy and peaceful in mind when I went onstage. That way, when the crowd got energized and started shouting—as crowds at rallies tend to do—I could resist doing the normal thing, which is to shout back. Men get to shout back to their heart's content but not women. Okay, I told this expert, I'm game to try. But out of curiosity, could you give me an example of a woman in public life who has pulled this off successfully—who has met the energy of a crowd while keeping her voice soft and low? He could not."

After all the speech training, Clinton was left baffled. "I'm not sure how to solve all this," she wrote. "My gender is my gender. My voice is my voice. To quote Secretary of Labor Frances Perkins, the first woman to serve in the US Cabinet, under FDR, 'The accusation that I'm a woman is incontrovertible.' Other women will run for President, and they will be women, and they will have women's voices. Maybe that will be less unusual by then."

Canada's first female prime minister, Kim Campbell, had an impassioned ad-lib oratorical style. As a result, she was ridiculed in the press for being arrogant, emotional, whiny, hectoring, angry, and shrill. Campbell's "verbosity is her Achilles heel," opined the *Vancouver Sun* in 1993. "She will fall back into the strident voice . . .

or even into the shrill, scolding voice," lamented the *Globe and Mail*. The most damning pronouncement of all came from the *Montreal Gazette*: Campbell "sounded like a woman." During the next election, Campbell's speaking was discussed in the Canadian media nearly four times as frequently at that of her male opponent, Jean Chrétien (146 remarks about Campbell, 38 for Chrétien).

Yet when Campbell chose to conform to the gendered expectations of well-behaving women, reading from a script, she was called inauthentic and boring. Her new approach was long-winded, lecturing, and "devoid of passion, poetry, or partisan fire," according to the *Vancouver Sun*.

Research by University of Alberta political science professor Linda Trimble has shown that many people find a woman speaking passionately to be scary (perhaps the audience fears she is skyrocketing into the outer stratosphere on a wildly unruly hormonal rant), whereas men doing the same are considered confident and statesmanlike. A 2007 study, "Gender Stereotyping of Political Candidates," found that male audiences rank men's speech as more knowledgeable, trustworthy, and convincing than women's speech, even when *they were saying exactly the same thing*.

"DID YOU JUST SHUSH ME LIKE A CHILD?"

One way to stop the shouting, chalk-screeching-on-chalkboard, nagging voices of women is to shush them. While we can well understand that hidebound ancient Romans and stuffy Victorian novelists wanted to silence women's voices, it is rather shocking that in February 2017 Senator Elizabeth Warren was officially shushed in the US Senate and excluded from debate during the confirmation hearings for Senator Jeff Sessions as Attorney General.

Warren had attempted to read out a 1986 letter by Coretta Scott King denouncing Sessions's nomination as a federal judge. King wrote, "Mr. Sessions has used the awesome power of his office to chill the free exercise of the vote by black citizens." Presiding Senate Chair Steve Daines of Montana interrupted Warren, citing Senate Rule XIX, which forbids imputing "to another senator or to other senators any conduct or motive unworthy or unbecoming a senator." Though the letter had already been accepted into the Senate Record thirty years earlier, the Senate voted along party lines to muzzle the senator. Daines told her to take her seat.

The arcane procedural rule that silenced Warren did not stop Bernie Sanders, Jeff Merkley of Oregon, Tom Udall of New Mexico, and Sherrod Brown of Ohio (all men, in case you didn't notice) reading out exactly the same letter without a single command to sit down and shut the hell up. This was clear-cut evidence of deeply embedded mechanisms to silence women and women alone. If a scold's bridle had been available to Mitch McConnell, we can well imagine him trying to padlock it onto Warren's head. Moreover, according to the same rule, McConnell should have silenced at least two men on the floor: Senator Ted Cruz in 2015 for calling McConnell a "liar" several times, and Senator Tom Cotton in 2016 for deriding Harry Reid's "cancerous leadership" as former Senate majority leader.

Four months after Senator Warren's high-profile shushing, during a session of the Senate Select Committee on Intelligence, Senator Kamala Harris insisted repeatedly that Deputy Attorney

General Rod Rosenstein provide a clear response to her questions instead of dodging, ignoring, and talking over them. The committee chairman, Senator Richard Burr, interrupted her, lectured her on "courtesy," and allowed Rosenstein to ramble as much as he wanted, using up valuable time in Harris's allotted five minutes of questioning.

When newly minted British MP Jess Phillips entered Parliament for the first time in 2015, she immediately knew something was wrong. In the lobby that leads up to the committee rooms of Parliament, she saw ornate carvings depicting both men and women as gargoyles. "The men are depicted open-mouthed in speech," she recalled in her memoir. "The women meanwhile are gagged, their mouths literally covered with stone muzzles. I have no idea if this was a less than subtle comment on gender equality by the architects of yesteryear, or a helpful suggestion. I hope it was the former, but sometimes I have my doubts."

Phillips is perhaps best known for guffawing when a male colleague suggested creating International Men's Day. "You'll have to excuse me for laughing," she said, a statement that quickly went viral. "As the only woman on this committee, it seems like every day to me is International Men's Day."

Phillips's male colleagues would certainly like to silence her, as she is one of the most vocal women in Parliament; her opinions are not only undeniably strong, but they cause—at least among women—a certain amount of falling on the floor rolling around laughing. However, since shoving a gag into her mouth would not go down well on the nightly news, men shush her instead. "If I am getting aggravated or am heckling in a debate," Phillips wrote, "I have noticed men from the opposition benches, men who shout and holler all they like, shushing me like I was a five-year-old . . . I am *not* a child; do not shush me."

She recalled one debate when a minister on the front bench shushed her. "You're not my dad," she said. "Don't you dare shush me while the men shouting around me get no such treatment! I say to you, good sir, you can take your shushing shushes and stick them up your shushing arse!"

Her advice to other women? "If anyone ever shushes you, my advice is to call it out. Ask the man in question, 'Did you just shush me like a child?' They will then be forced to verbalize their dislike of your opposition to their views and will fall apart almost instantly."

If shushing doesn't work, there's the male fantasy of physically preventing women from speaking, a kind of twenty-first century virtual scold's bridle. On May 22, 2020, Donald Trump retweeted a photoshopped image of House Speaker Nancy Pelosi with a ragged piece of duct tape clapped over her mouth. Clinton has been shown in memes and cartoons with a closed zipper for lips, a muzzle on her mouth, and a box with air holes over her head. Nor was such overt silencing of female politicians limited to the virulent Trump-era politics of the United States. In October 2016, a political candidate tweeted a request to silence First Minister of Scotland Nicola Sturgeon by taping her mouth shut.

Francesca Donner, formerly the gender director of the *New York Times*, finds it extremely unfortunate that, due to the necessity of masking to stop the spread of COVID-19, when President Joe Biden speaks at official White House events, Kamala Harris stands supportively behind and to the side of him, her mouth swathed in fabric. "Decades from now," Donner lamented in an interview for this book, "when COVID is a vague memory, people will see the nation's first female vice president wearing what looks like a muzzle."

DROP THE GILLARD TWANG

A week before she became prime minister in 2010, Deputy Prime Minister Julia Gillard was voted as having the most impressive voice in Australian politics, according to a survey of some 3,600 respondents conducted by the *Age* newspaper. Curiously, she won against the reassuringly deep tones of several male colleagues.

Once she landed the top governmental position, however, her voice seemed to become more annoying. As speech expert Dean Frenkel wrote in the *Sydney Morning Herald*, "Sound characteristics are unique in the way they grow on listeners. When prime ministers' voices are media-exposed beyond saturation point, once forgettable and minor irritating characteristics become magnified and incrementally annoying. When these kinks are repeated every night, they become more and more apparent."

In other words, it wasn't that Gillard's voice was annoying; it's just that people were hearing it too often. While admitting that most of the prime minister's speaking skills were excellent, Frenkel suggested she lose the "Gillard twang"—a regional accent of which she was quite proud—which resulted in "heaviness of tone and clumsy treatment of vowels." There was too much gravity in her voice, he said—though we might imagine that prime ministers *should* have gravity in their speaking. To correct her oversaturation of gravity, she should "tap into a greater range of melody and more frequent higher melody. This would raise her energy and sound more natural." Simply put, her natural voice didn't sound natural, and if she practiced a great deal, she would sound more natural.

She should be "more resonantly balanced between her throat and mouth," which sounds both baffling and painful, and should improve her "heavy and earthy" tones with "lighter and brighter tones that introduce more melodious qualities." (Why does this sound like wine tasting? *Bright and flamboyant, opulently oaked!*) "Vocally, she seems to have little experience of singing" (and, as we know, a successful operatic career is a prerequisite to political office for men and women alike). To solve these serious and disturbing problems with her voice, the prime minister should take singing lessons, "but not the gung-ho footy anthem type." We're not sure what types of songs the author had in mind, but perhaps some light arias would do the trick. That is, if she could find the time what with dealing with devastating fires, floods, cyclones, unemployment, and international relations and generally running the country.

It is likely that men don't like women's voices because deep down there's the feeling women shouldn't be talking to begin with. And as deputy prime minister, second-in-command to a man, Gillard's voice was less irritating than when she was raised to the top job. Because the problem was not her voice. It was her position. And it's much easier to say we don't like a particular voice than acknowledging we don't like the fact that a woman is talking. Or even expressing her voice through writing.

In 2016, British journalist Caroline Criado Perez received a flood of online abuse because of her advocacy for feminist causes. While many women in her situation would have hammered on the block button as soon as each foul post emerged—in a kind of sexist abuse whack-a-mole—Criado Perez studied them, curious to find a common denominator. "Thousands of threats I received . . . focused on my mouth, my throat, my speech," she reported. "The message was simple and clear: these men very much wanted me to stop talking." Even though she was writing.

In 2017, Hillary Clinton published *What Happened,* her memoir of the 2016 presidential election. Was the nation remotely curious about the personal experience of the woman at the center of the most autopsied election ever? Not much. The New York *Daily News* advised, "Hey Hillary Clinton, shut the fuck up and go away." But even left-leaning publications wanted her to gag herself. *Vanity Fair* trumpeted the headline, "Can Hillary Clinton please go quietly into the night?" and the *New York Times* asked, "What's to be done about Hillary Clinton, the woman who won't go away?"

When another unsuccessful 2016 candidate, Bernie Sanders, published his memoir of the election, no one told him to muzzle himself, and he wasn't even the nominee. Similarly, when failed 2008 Republican presidential candidate Mitt Romney wrote a blueprint for American greatness, *No Apology,* in 2010, not a single person suggested he take a spaceship to Alpha Centauri and stay there.

As the feminist author Kate Manne wrote in her book *Down Girl: The Logic of Misogyny,* "When a woman competes for unprecedented high positions of male-dominated leadership or authority, particularly at the expense of an actual male rival, people tend to be biased in his favor, toward him. That is, there will be a general tendency, all else being equal, to be on his side, willing him to power, and this in turn predictably leads to biases against her. So when she speaks against or over him, by disagreeing with him, interrupting him, laughing at his expense, or declaring victory over him—it would be natural for her voice to be heard as grating, raspy, shrill, or otherwise painful sounding. We do not want to hear her say a word against him, so she becomes hard to listen to."

In the 1920s, researchers at Bell Laboratories, conducting studies on the transmission of human voices over the telephone, found that women's voices were generally as loud as men's, but they were more difficult to understand. They wondered if there was a problem in transmitting higher frequencies but found that listeners could hear the high-pitched sound of a flute perfectly well. Which leaves us to ponder whether women's voices were harder to decipher because people on the other end of the phone—both men and women—just weren't listening.

"MR. VICE PRESIDENT, I'M SPEAKING"

Another time-tested stratagem to silence women is to simply roll over them. At the traditional opening of the Supreme Court term, the first Monday in October 1981, Sandra Day O'Connor, the newly minted first female justice, took her place at the end of the long mahogany bench above where the lawyers would argue. As the first case was presented, the other justices immediately began firing questions at the lawyer standing at the lectern ten feet away. For thirty minutes, as the legal arguments and questions flew back and forth in a complex case involving oil leasing, she remained silent. As she reported in her journal that night, she wondered, "Shall I ask my first question? I knew the press is waiting—All are poised to hear me." From her seat on the high bench, she began to ask a question, but almost immediately the lawyer talked over her. "He is loud and harsh," she wrote, "and says he wants to finish what he is saying."

In a 2017 study conducted by Northwestern University's Pritzker School of Law, researchers found that female Supreme Court justices were interrupted roughly three times as often as their male counterparts. For instance, in 1990, when O'Connor was the only woman on the nine-member court, 35.7 percent of all interruptions occurred when she was speaking. In 2002, 45.3 percent of the interruptions were directed at the two female justices, O'Connor and Ruth Bader Ginsburg. In 2015, 65.9 percent of all interruptions were directed at the three female justices (Ginsburg, Sonia Sotomayor, and Elena Kagan). If there was ever an all-female US Supreme Court, we can be fairly certain that not one of the justices would ever get the opportunity to talk in the presence of male lawyers.

As the first female president of the Philippines, Corazon Aquino, modest and polite though she was, soon had enough of men rolling over her speaking. In November 1986, just nine months into her first term of office, she said, "One distinct quality I have observed in the men who would discount my abilities, diminish my role, or who cannot bring themselves to imagine that I shall rule this country for the entire term of a presidency, is their ability to out-talk me at every opportunity." She had noticed "a crop of garrulous men with better and brighter ideas on how to run my government or what I should do with myself. . . . I would like to think that I have managed to have the last word and the last task of having to set things back in order after these men were finished."

In the October 7, 2020, vice presidential debate, Kamala Harris faced off against incumbent vice president Mike Pence. Although the debate moderator, Susan Page of *USA Today*, did her best to corral Pence into his allotted time, her efforts were mostly fruitless. Not only did he avoid answering many of her questions, he interrupted Harris's answers to the debate questions no less than ten times.

"Mr. Vice President, I'm speaking," Harris asserted again and again. Twitter users cheered her for refusing to let the man steamroll her.

In addition to constantly interrupting her, Pence mansplained her, a white man lecturing a Black woman on why Black people in the US do not suffer from systemic racism. He offered his venerable wisdom and guidance on the interaction of law enforcement with the Black community, an area of Harris's expertise going back many years as district attorney.

"I will not sit here and be lectured by the vice president on what it means to enforce the laws of our country," Harris said. "I'm the only one on this stage who has personally prosecuted everything from child sexual assault to homicide."

Harris had a fine line to tread during that debate. Too fierce and she would have looked angry, hormonal, unprofessional. Too deferential and she would have looked weak, unsuited for the job. She managed to hover in the sweet spot.

"No advanced step taken by women has been so bitterly contested as that of speaking in public," said Susan B. Anthony, the pioneering women's rights activist, at the end of her long career. "For nothing which they have attempted, not even to secure the suffrage, have they been so abused, condemned and antagonized."

No one—from Homer to Saint Paul to Donald Trump—wants to hear a powerful woman speak. Because the Voice of Authority is always male.

CHAPTER 6

THE MYSTERIOUS UNLIKABILITY OF FEMALE CANDIDATES

*The more successful and therefore ambitious a woman is,
the less likable she becomes.*

—Hillary Clinton, 2017 Women in the World summit

On October 7, 2020, Senator Kamala Harris won the vice presidential debate, according to FiveThirtyEight.com, with 69 percent of respondents judging her performance as good, compared to 60 percent who felt the same about Vice President Mike Pence. Similarly, 61 percent approved of her policies, while only 44 percent approved of his.

That same evening, Iowa senator Chuck Grassley tweeted that he would prefer to have dinner with Pence, who was far "MORE LIKABLE," than dine with Harris. And Fox News host Bret Baier asked former White House deputy chief of staff Karl Rove if he found Harris "likable."

Rove replied, "If she was trying to, she failed at it."

The next day, President Trump called in to Fox Business to opine that Harris was "totally unlikable" and later called her "this monster that was on stage with Mike Pence."

Alas, Kamala Harris isn't the only female candidate who has been called "unlikable." A 2018 *New York Times* poll found that female candidates in general are just more unlikable than males. According to research from the Barbara Lee Family Foundation, when it comes to unlikability, voters "have an 'I know it when I see it'" mindset. And when they see it, it is almost always a female candidate.

The thing about a candidate's unlikability is that you can't quite put your finger on what it is that bothers you. It's like a faint, revolting odor wafting in front of your nose, and just before you identify it, *poof!* It's gone.

And the intangibility seems to be the point. Because the thing people just can't quite seem to put their finger on may be their own misogyny. The likability issue is used by those who are sexist— including a great many women—and simply can't admit it to themselves or anyone else.

Kim Campbell, the first female prime minister of Canada for a little more than five months in 1993, has an explanation for the unlikability of powerful female politicians. "Everything you do is judged at the highest possible scrutiny so people can validate their own sense of discomfort that you're there," she told the *Toronto Star* in 2020.

Those of us who grew up imbibing the unspoken but omnipresent tenets of the Patriarchy—which is pretty much all of us—are naturally more comfortable with powerful men than with powerful women. Sociologist Marianne Cooper wrote in a 2013 *Harvard Business Review* article that "the very success" of powerful women, "and specifically the behaviors that created that success—violates our expectations about how women are supposed to behave." As a society, Cooper wrote, "We are deeply uncomfortable with powerful women. In fact, we don't often really like them. . . . Women are expected to be nice, warm, friendly, and nurturing. Thus, if a woman acts assertively or competitively, if she pushes her team to perform, if she exhibits decisive and forceful leadership, she is deviating from the social script that dictates how she 'should' behave. By violating beliefs about what women are like, a successful woman elicits pushback from others for being insufficiently feminine and too masculine."

How, then, are women politicians to showcase their accomplishments without seeming unfeminine and immodest? How should they handle political criticism and remain fairly likable? In June 2021, the Barbara Lee Family Foundation released the study "Staying Power," which examines strategies for women incumbents running for reelection. "They will not assume she is doing a good job, nor will they simply take her word for it," said Executive Director Amanda Hunter in an interview for this book. "That's why it is particularly important for a woman incumbent to establish her record as a leader. Likewise, when opponents critique a woman candidate for the job she is doing, voters look for her to defend her record by delivering a decisive response. Beyond addressing the issues in question, voters want to see a woman incumbent handle criticism with strength and composure."

Despite strength and composure in the face of vicious criticism, Hillary Clinton has had to wrestle with the likability issue more than perhaps any other politician. "Everybody has a different personality, a different temperament, a different public persona," she said in an interview for the book *Women and Leadership*, "so you can like or dislike people for whatever reason. But women are much more likely to be judged unlikable if they're assertive, if they're strong, if they are willing to stand up and speak out. I saw it over and over and over again in my campaign. People would say, 'There's something about her I don't like.' Then, when pressed on what it was, they could not provide any more detail. They would say, 'I don't know. It's just something, I don't know.'"

At least radio host Rush Limbaugh had concrete reasons for not liking Clinton. It was because she was "not soft and cuddly, not sympathetic. Not understanding." (Not likely to make him a sandwich.)

During her two presidential campaigns, Clinton's thoughtful poise was perceived as cold, plastic, and machinelike. This was proven when a fly landed on her eyebrow for a split second during a presidential debate and she ignored it, evidence that she was, in fact, a robot. (Though no one said Mike Pence was a cold, plastic machine when a black fly landed on his silver-white head during the 2020 vice presidential debate and stayed there for two minutes and three seconds, but who's counting?)

Another female leader deemed to be a robot was UK prime minister Theresa May. Hardworking, reserved, and calm, she was known as "unclubbable," a term that indicates she kept to herself rather than drinking pints and backslapping with her colleagues at the local pub. She spoke in sensible, measured tones. She did not screech in passion or gesticulate wildly (and just imagine the criticism if she had). Perhaps because of these attributes, she found herself christened the "Maybot" by John Crace, a writer for the *Guardian*. Crace even wrote a book about May called *I, Maybot: The Rise and Fall*. In July 2017, the *New Statesman* published a cartoon of May as a robot with the headline "The Maybot Malfunctions." According to the *Financial Times*, a Maybot was "a prime minister so lacking in human features that she soon required a system reboot."

New Zealand prime minister Helen Clark was also seen as robotic. One reporter described her as "a cold, unemotional, purpose-driven woman with a steely determination to succeed against the odds." (Which is actually a good thing in a prime minister, right?)

Yet those women who do react in a non-robotic manner, whose faces come alive with humor or disbelief, also get criticized. Senator Elizabeth Warren, who waved her arms and wagged her finger in the 2020 Democratic primaries in a way no robot could, has often been described in the press as "hectoring," "school marmish," and "angry." When Bernie Sanders did the same, he was praised for his passion and authenticity.

Former Australian prime minister Julia Gillard said in an interview for a doctoral dissertation, "You hold yourself back from getting too angry, too animated, too passionate because you're fearful of being labelled as hysterical or shrill. You end up walking quite a narrow behavioral pathway and I think it's no mystery that women leaders are often therefore described as 'aloof', 'robotic', 'cold'."

During the 2020 vice presidential debate, Kamala Harris's non-robotic facial expressions clearly showed what she was thinking as Mike Pence spoke, thoughts such as *You've got to be kidding me.* Conservative commentator Bill O'Reilly said, "Senator Harris is articulate but comes across as arrogant and the facial expressions are hurting her." One tweeter wrote, "Kamala's face when Pence is talking is the same face my mother made whenever I was explaining why I didn't make curfew." Republican candidate David Dudenhoefer, who was running for Congresswoman Rashida Tlaib's seat in Michigan's 13th congressional district, wrote, "Kamala Harris is unlikable with her smug facial expressions."

It's rare for men to be called unlikable. And it probably really wouldn't matter if they were. The Barbara Lee Family Foundation discovered that voters are more comfortable voting for men they don't like—Donald Trump being a prime example. Many of his most ardent supporters really couldn't stand the guy personally—they hated his nasty tweets and unending boasting and spewing of lies—but they approved of his policy initiatives and would beat tracks to the polls on Election Day.

"When women seek executive office," said Amanda Hunter of the Barbara Lee Family Foundation in an interview for this book, "they have to really satisfy both gender stereotypes: show they are strong enough to lead, but they also have to maintain a level of femininity to keep their likability, which is nonnegotiable. Voters will not vote for a woman they do not like.

"Trump's confrontational manner would never be permitted in a woman."

It's not that women *can't* be likable. It's just that it's a difficult balancing act. On a tightrope. Over the Grand Canyon. Juggling fifteen balls. With a full tea tray on her head.

So how do female candidates become more likable?

Well, for one thing, they should never boast of their accomplishments. During the 2012 Senate race, Democratic analyst Dan Payne advised Elizabeth Warren "to show a little modesty." Political speech coach Chris Jahnke told her female clients that voters would find them more likable if they shared credit for their accomplishments, deflecting praise to their talented team.

What about humor? That ought to make them likable. Everyone likes to laugh. Not so fast. In a 2019 study, "Gender and the Evaluation of Humor at Work" in the *Journal of Applied Psychology*, researchers had a man and a woman give two versions of a speech to employees. When the man cracked a few jokes, his audience gave him top marks for leadership, status, and performance. When the woman used the exact same humor, they rated her lower across the board.

In an article in the *Washington Post*, Alyssa Mastromonaco, former deputy chief of staff in the Obama White House, said that every time female politicians "display humor, they're called inauthentic or they're trying too hard." But if they don't display humor, they are cold, self-conscious, and robotic. Media coach John Neffinger warned, "The most common danger is a female candidate using self-deprecating humor to project warmth and totally undercutting her strength." And that would make her unpresidential. Lacking in substance.

Female politicians need to make sure they don't come off as nagging or like a schoolteacher or librarian. In January 2020, one journalist wrote that Senator Amy Klobuchar was like "a mean-spirited elementary-school librarian who is about to remind us for the fifth time to use our indoor voices." CNN's Jack Cafferty called Hillary Clinton a "scolding mother, talking down to a child." (Here we go again, back to the Jungian archetype of women as Nurse Ratched.)

The words "inauthentic" or "phony" are almost always used to describe female candidates, just as "unlikable" is. Soon after Joe Biden's announcement of his running mate, Trump's reelection campaign manager Bill Stepien called Kamala Harris "phony" for criticizing Biden during the presidential debates and then agreeing to join his ticket (something, of course, no male politician would ever do). "She has an air of inauthenticity," said Cliff Sims, who wrote speeches for the 2020 Republican National Convention, "which is a major problem at a time when plastic politicians just aren't connecting with voters. That's why the 'phony' line of attack really hits. It rings true."

Ah, yes, that air of inauthenticity.

Elizabeth Warren, too, has had her share of the same criticism. When, at a campaign rally in January 2020, she jumped around the stage for a few moments to the song "Respect," trolls attacked her for inauthenticity.

She was inauthentically *dancing.*

Christina Reynolds, vice president of communications for EMILY's List, told *U.S. News* that likability "is a cudgel used against women. Women are supposed to be likeable, and that's supposed to be a part of their authenticity. Whereas men can be authentic without being likeable."

Hillary Clinton, too, was accused of being phony and inauthentic for the same reason she was dubbed robotic. When asked a question, she had the habit of pausing to consider before answering. "I learned to 'think before I speak,'" she wrote, and "sometimes sound careful with my words. It's not that I'm hiding something, it's just that I'm careful with my words." In 2016, Chuck Todd, then host of *Meet the Press*, actually criticized Hillary Clinton for being "overprepared" for her first debate against Donald Trump, leaving us to scratch our heads wondering how a candidate can be overprepared for a presidential debate and if such a statement would ever have been made about a male candidate.

After being condemned for not smiling enough, Clinton made an effort to smile more during her first presidential debate in 2008 against Mitt Romney. But many commentators said she smiled *too* much. And that made her look inauthentic.

She was inauthentically *smiling*.

"Again I wonder what it is about me that mystifies people," she wrote, "when there are so many men in politics who are far less known, scrutinized, interviewed, photographed, and tested. Yet they're asked so much less frequently to open up, reveal themselves, prove that they're real." Clinton pointed out that Barack Obama was just as controlled as she was. "He speaks with a great deal of care, takes his time, weighs his words. This is generally and correctly taken as evidence of his intellectual heft and rigor. He's a serious person talking about serious things. So am I. And yet, for me, it's often experienced as a negative. I have come to terms with the fact that a lot of people—millions and millions of people—decided they just didn't like me," Clinton concluded. "Imagine what that feels like."

In 2013, Prime Minister Julia Gillard posed for *Australian Women's Weekly* sitting in an armchair knitting a stuffed kangaroo for the baby of the Duke and Duchess of Cambridge. You might think that such a feminine hobby—one that she had enjoyed for years—would evince a warm smile from the Patriarchy. No such luck. The Australian media called it "contrived," "a hobby synonymous with mad old aunts," and "a bit of a stunt" that showed "a lack of connection" with the public. One politician from an opposing party told reporters, "We know the prime minister is good at spinning a yarn, now we have a picture to prove it."

She was inauthentically *knitting*.

The deputy editor of *Australian Women's Weekly* wrote, "The federal budget is $19 million in deficit. And what's she doing? She's knitting." Clearly, if Gillard had had an ounce of concern for the nation, she would have been printing money.

Gillard's supporters noted that Tony Abbott, the opposition leader at the time and subsequently prime minister, was often shown pursuing his favorite hobbies—cycling and surfing—without being excoriated in the press, despite the rising national deficit.

"SMILE!"

One way that women can be more likable is to smile. (If they don't, it's a good bet that men will tell them to, whether they are running for president or walking down the street.) In October 2015, Republican presidential candidate Carly Fiorina said that she had been told she hadn't smiled enough during the last debate.

A 2005 *Seattle Times* column reported that Senator Maria Cantwell of Washington State was called "Maria Cant-Smile" because of her "serious, almost cold, personal demeanor."

When former UN ambassador Susan Rice was in the running for Joe Biden's VP pick, former Democratic Party chair Ed Rendell told the *Washington Post*, "She was smiling on TV, something that she doesn't do all that readily. She was actually somewhat charming on TV, something that she has not seemed to care about in the past." Which is a backhanded compliment if ever there was one.

Hard as it is to believe, sometimes even women are guilty of telling other women to smile. In 2019, then White House press secretary Sarah Huckabee Sanders said then House minority leader Nancy Pelosi "should smile a lot more."

After Hillary Clinton won several primary victories in March 2016, NBC's Joe Scarborough tweeted at her, "Smile. You just had a big night." When CNN's Anderson Cooper asked her if she thought the suggestion was sexist, she replied, "Well, let me say, I don't hear anybody say that about men. And I've seen a lot of male candidates who don't smile very much and who talk pretty loud. So I guess I'll just leave it at that."

Clinton had a point. If a male leader is serious and thoughtful, would anyone tell *him* to smile? To try harder to be charming? Why didn't anyone tell Donald Trump to be charming? (Maybe it would have made a difference?)

And why do these people think they can tell us what to do with our body parts? They might as well insist that we hop on one foot and clap our hands upon command.

It is probably about control. Men have always told us how we should look so they must think it's okay to tell us what to do with our faces. Though most likely this doesn't often rise to the level of a man's conscious thought, it is clear that a smiling woman would be far more likely than an unsmiling one to pour him a drink, massage his feet, strip off her clothes, and jump into his arms. Perhaps, though, it's more than that, and deep down in the subconscious, a smiling woman confirms the Patriarchy. The Lord and Master is on his throne; women are happy with their assigned place, and all is in Divine Order. Women like Sarah Huckabee Sanders who tell other women to smile are merrily jumping on the patriarchal bandwagon, where they can expect a gleeful welcome and collegial pat on the back.

Those women who fail to smile dazzlingly at strange men when walking down the street because they are thinking about work, bills, and dying friends are not just unlikable; they are often accused of having "resting bitch face"—which means the individual so described is nasty, combative, and sour. This term is clearly sexist because it is only ever said about a woman (with the possible exception of Kanye West). There is no such a thing as "resting asshole face."

"LET THEM EAT CAKE"

Some famous women in history were disliked because of their supposed spendthrift frivolity. Cleopatra, for instance. As evidence of her profligacy, according to Roman writers who loathed her, Cleopatra made a bet with the visiting general Mark Antony that she could spend ten million sesterces on a single dinner party, a sum equivalent to about $20 million in today's money (enough to buy a lavish Mediterranean villa in Cleopatra's day). She removed an enormous pearl from one of her earrings—the Roman author Pliny called it "the largest in the whole of history," a "remarkable and truly unique work of nature"—and dropped it into a cup of vinegar where it sizzled like an Alka-Seltzer. Once it had dissolved completely, she drank it, shocking Antony and her guests.

The problem is, pearls don't dissolve in vinegar unless they are powdered, or are put in very hot, highly concentrated vinegar for longer than a dinner party. Either the queen played a clever trick on her Roman visitors to dazzle them with her astonishing wealth, or later writers made up the tale out of whole cloth to show what a dissipated, frivolous fool she was, the kind of woman whose country really was in better hands as a Roman province ruled by supremely sexist men.

Now let's consider Marie Antoinette. Sometimes she escaped the rigid etiquette and dizzying décor of the Palace of Versailles by going to the Petit Trianon—a mini-mansion on the palace grounds, a place of comparative simplicity. The previous king, Louis XV, had built it for his mistress, Madame de Pompadour, but the underground press reported that Marie had had it built for herself at great cost, plastering the walls with gold and diamonds.

When her brother Joseph, emperor of Austria, visited her in 1781, rumors abounded that she had given him trunks full of gold to spirit out of France and take back to Vienna. The pamphlets portrayed her as a drunk when, in fact, she was one of the few individuals at court who never touched a drop of alcohol.

During the famines of the 1780s, when the people complained they had no bread, Marie Antoinette is famous for saying, "Let them eat cake." Which never happened. The comment had been around for at least a century previously, when Queen Marie Thérèse, a Spanish princess married to Louis XIV, supposedly said, "If there is no bread, let the people eat the crust of the pâté" (which admittedly does not have the same ring to it). In 1751, Louis XV's daughter Sophie referred to the old saying when her brother reported hearing cries of "Bread! Bread!" on a visit to Paris. The philosopher Jean-Jacques Rousseau wrote in 1765 that a long-ago great princess had said "Let them eat cake." But the phrase has stuck as having been uttered by Marie as proof of her selfish, frivolous disregard of the people's hardships.

LIAR/UNTRUSTWORTHY/BACKSTABBER

Powerful women are often viewed as untrustworthy for some reason. Perhaps they are suspicious by virtue of being powerful, a clear violation of the Patriarchy. Untrustworthiness, rather like likability, is vague, hard to put one's finger on, a gut feeling unsubstantiated by fact. She's up to something (Fraud? Conniving? Nude pole dancing?) and if we don't know yet exactly what it is, we'll surely find out later.

In a parliamentary system, a leadership challenge is an accepted process for a party to determine whether it wants to replace an incumbent leader. And yet when a woman unseats a man, accusations of backstabbing abound. For instance, in 1993, New Zealand deputy leader of the opposition Helen Clark and her supporters from the Labour Party were unhappy with the party's poor election results and concerned that it would not do well in the 1996 election. Clark decided to try for the leadership spot, a position held by her boss, Mike Moore. She requested a leadership ballot, where members of her party in Parliament would choose either her or Moore. Moore could have yielded his position, but decided to stay and fight, a battle he lost.

Though Clark's challenge was exactly what many male politicians had done for decades, the media acted as if she had physically harmed Moore. She had "betrayed" him, "conspired against" him. She was stealthy, untrustworthy, mean-spirited, heartless, and power-hungry. In her first interview after winning the vote, she was introduced with this teaser: "Next, in the studio with Mike Moore's blood still fresh on her hands, the new leader of the Labour Party, Helen Clark." The interviewer asked, "Helen Clark, I can't see any blood on your hands, but what's it like to knife a leader in the back like that?" Moore was represented as the undeserving victim of Clark's unruly political ambitions. "Given the circumstances of this leadership coup," the interviewer demanded, how can we trust you?

In 2010, Julia Gillard became Australia's first female prime minister in a similar manner. She had served Prime Minister Kevin Rudd for three years as his deputy prime minister, often trying to smooth over the chaos caused by Rudd's unfocused management style, his lateness to meetings, and his frequent bewilderment at what was going on. While the public didn't know of the turmoil going on behind closed government doors, they increasingly disapproved of his blunders on climate change and a mining tax. Gillard believed she would make a better leader and had the votes in her party's caucus to get the job. When she told Rudd she would challenge him, he held a news conference announcing he would fight to stay in his job. Then, realizing he did not have the confidence of the party, he agreed to step down, and Gillard was elected unopposed.

News reports accused Gillard of "killing" and "decapitating" Rudd, of engineering an "assassination," of deciding "to execute Rudd politically." "When the opportunity came, the ambitious Gillard did not hesitate to take up the knife and plant it in Rudd's back," the *Courier-Mail* trumpeted. The *Age* tsk-tsked, "Certainly, anyone expecting Parliament to be a softer, gentler place because a woman is in charge is likely to be disappointed." Another article in the *Age* with the headline "Nice Girls Don't Carry Knives" opined, "So Julia Gillard, who has arrived in the prime ministership with the image of the clean, fair player, knows she has to be persuasive in explaining how she came to plunge one into Kevin's neck. So she has a mantra. She had to get the government 'back on track.'"

Just two months after she became prime minister, Gillard called a national election as a means of legitimizing her position among voters. Yet the election did not give her a majority, and she had to negotiate agreements with the Greens and some Independents to form a government. The Greens demanded she set a price on carbon, which she had resolutely promised never to do during her campaign. Given the new circumstances, she instituted the carbon tax. A pundit dubbed her "Juliar," a name that quickly went viral.

Such a change of position is fairly common in politics. Many of

Gillard's male predecessors had found themselves in similar positions. Paul Keating, for instance, had promised to reduce taxes in 1993 and found himself unable to do so. Keating, however, while criticized for flip-flopping, was never called "Pauliar." John Howard, prime minister from 1996 to 2007, swore in 1995 he would "never, ever" institute a GST, a goods and services tax. In 1998, when he found it necessary to reverse course, the media and the public duly chided him, but not with the vicious hatred lobbed at Gillard. In March 2013, Kevin Rudd stated that he wanted to make it "100 percent clear to all members of the parliamentary Labor Party" that he would *never*, under any condition, take up party leadership again. Yet only three months later he did so, winning back his position as prime minister. He was not called a liar.

Women have a great advantage in rising to the top of a parliamentary system as opposed to a presidential one. Margaret Thatcher and Theresa May of Great Britain, Golda Meir of Israel, Indira Gandhi of India, Benazir Bhutto of Pakistan, Julia Gillard of Australia, Angela Merkel of Germany, and many other female national leaders operated within parliamentary systems. They were chosen not by the public in a general election, but by their colleagues, other publicly elected officials, who have known and worked with them for years. These few hundred electors understand their skill set with regard to public speaking, foreign and domestic affairs, compromising, working within a budget, and passing legislation.

Women vying for the top executive position in a parliamentary government don't necessarily need to be warm and fuzzy to the public at large. And ambition is seen as an asset by her political colleagues rather than a character flaw by the populace. Had Hillary Clinton served in a parliamentary system, given her experience and qualifications, she likely would have become prime minister.

Looking back on her 2016 run, Clinton said, "It's not easy to be a woman in politics. That's an understatement. It can be excruciating, humiliating. The moment a woman steps forward and says, 'I'm running for office,' it begins: the analysis of her face, her body, her voice, her demeanor, the diminishment of her stature, her ideas, her accomplishments, her integrity. It can be unbelievably cruel. . . .

"In my experience, the balancing act women in politics have to master is challenging at every level, but it gets worse the higher you rise. If we're too soft, we're not cut out for the big leagues. If we work too hard, we're neglecting our families. If we put family first, we're not serious about the work. If we have a career but no children, there's something wrong with us, and vice versa. If we want to compete for a higher office, we're too ambitious. Can't we just be happy with what we have? Can't we leave the higher rungs of the ladder for men?"

PLAYING THE GENDER CARD

Perhaps the most unlikable thing a female can do is complain of sexist treatment. Such behavior often doubles the injuries sustained: first, the sexism itself, and second, the accusations of "playing the woman card" for special treatment, of being a whiner, a liar, a complainer, not being a team player, and having no sense of humor. A 2001 study published in *Feminism & Psychology* found that pointing out sexism made the women who did so "liked less" by men than when they sucked it up.

As Congresswoman Pat Schroeder wrote in her 1998 autobiography, "Professionally, we women are afraid to express anything less than perfect professional contentment for fear of hearing if we can't take it, go home. We bottle up our feelings, afraid of being labeled whiners."

Julia Gillard's most defining moment occurred on October 9, 2012, when she gave what became known as "the Misogyny Speech," probably the biggest public reproach of sexism ever recorded. Her chief political opponent, opposition leader Tony Abbott, had made a motion to have Peter Slipper, the Speaker of the House, removed from office after it became known that he had sent sexist texts to his aide. Abbott said that every day Prime Minister Gillard supported Slipper was "another day of shame for a government which should already have died of shame" and accused her of sexism.

Gillard had been the victim of vicious sexism for more than two years as prime minister, and for many years before that in her political career, much of it from Abbott himself, without ever uttering a word of complaint. Now, to be accused of sexism herself, by the very man who had launched a misogynistic crusade against her, was too much to bear. "I do not normally think in swear words," she recalled in her memoir, "but my mind was shouting, 'For fuck's sake, after all the shit I have put up with, now I have to listen to Abbott lecturing me on sexism. For fuck's sake!'"

She rose to her feet and launched into a fifteen-minute speech— referring only briefly to notes she had made of his most appalling sexist comments over the years—that left many of her political opponents and supporters alike gaping in disbelief. She said, in ringing tones, "Thank you very much, Deputy Speaker, and I rise to oppose the motion moved by the Leader of the Opposition. And in so doing I say to the Leader of the Opposition I will not be lectured about sexism and misogyny by this man. I will not. And the Government will not be lectured about sexism and misogyny by this man. Not now, not ever."

She continued, "The Leader of the Opposition says that people who hold sexist views and who are misogynists are not appropriate for high office. Well, I hope the Leader of the Opposition has got a piece of paper and he is writing out his resignation. Because if he wants to know what misogyny looks like in modern Australia, he doesn't need a motion in the House of Representatives, he needs a mirror. That's what he needs."

She then called Abbott out for hypocrisy based on his own many misogynistic statements in the past, such as saying in an interview, "If it's true that men have more power generally speaking than women, is that a bad thing?" And "Abortion is the easy way out." And "What the housewives of Australia need to understand as they do the ironing . . ."

Gillard quipped, "Thank you for that painting of women's roles in modern Australia."

She recalled the time when Abbott said in a parliamentary discussion, "If the Prime Minister wants to, politically speaking, make an honest woman of herself . . ." as a reminder that she was not married to her partner. That, she said, was "something that would never have been said to any man sitting in this chair."

She added, "I was offended when the Leader of the Opposition went outside in the front of Parliament and stood next to a sign that said, 'Ditch the witch.' I was offended when the Leader of the Opposition stood next to a sign that described me as a man's bitch."

Gillard made clear that she was offended by Slipper's text messages—as she was offended by all sexism—and wanted to let the investigative process already underway play out. (Slipper would resign later that day.)

She ended with, "Good sense, common sense, proper process is what should rule this Parliament. That's what I believe is the path forward for this Parliament, not the kind of double standards and political game-playing imposed by the Leader of the Opposition now looking at his watch because apparently a woman's spoken too long."

And with that, she sat down to a stunned silence. Years later, she said, "Looking back, I think it was driven by a deep frustration that after every sexist thing directed at me that I'd bitten my lip on, now I was going to be accused of sexism—the unfairness of that. That anger propelled it."

Almost as soon as Gillard had finished speaking, tweets and links started whistling around the world. The speech was viewed more than one million times in the first week alone. The public— particularly the female public—celebrated. Her speech was voted the most unforgettable moment in the history of Australian television. But a study of 251 articles about the speech that appeared in major Australian publications in the week that followed found that most were negative. The same publications that rarely reported on Abbott's horrifyingly sexist comments made sure to attack Gillard for bringing them up. The *Courier-Mail* dismissed her speech as "unedifying" and "irrelevant," a "tawdry sideshow, brimming with confected outrage and affront. . . . Get over it, and instead get on with the business of delivering economic growth and stability; of actually delivering on big-ticket policy agenda items."

A columnist in the *Advertiser* wrote, "I never heard Thatcher scream out in the House of Commons that her critics were sexist misogynists. She would have thought that sounded weak. In the Thatcher school, any political leader who complained about her critics being mean and unfair was seen as unfit for the job. . . . This is dangerous territory for Julia Gillard. She is our national leader. She's our Prime Minister. We expect her to govern wisely and calmly and to dismiss her critics politely and gracefully. . . . She should rise above that." Somehow it is hard to imagine the writer suggesting a man suffering unrelenting personal attack for two years to govern wisely and calmly and dismiss his critics politely and gracefully.

The general feeling in the Australian press seemed to be that Gillard's pointing out misogyny was inexcusable, far worse than the misogyny itself. Why couldn't she ignore it, rise above it? In the Misogyny Speech, Gillard had spoken The Thing That Dare Not Speak Its Name.

The *Sydney Morning Herald* stated that the speech could cost her the next election (and some research indeed suggested that it did). The *Sunday Telegraph* said, "Playing the gender card is the pathetic last refuge of incompetents and everyone in the real world knows it." Her speech was described in the *Australian* as "an affront to women who *have* suffered harm from sexism and misogyny."

The *Australian* reported, "The notion of Gillard the student politician, full of pointless sound and fury, damages her more than

Abbott. . . . Abbott appears to be a normal guy from the suburbs with a mortgage and three daughters. He's not all that different from the strivers at the golf club or the blokes doing the barbecue at the netball." In other words, Abbott, despite his misogyny, was okay because he had a house, daughters, golf clubs, and he grilled burgers on the barbecue. Ironically, Gillard was accused of *starting a gender war*. She was painted as the villainous aggressor; Abbott as the hapless, harmless victim. By speaking out about misogyny, Gillard, an irrational, hormone-crazed woman, had ruined her party, her gender, the country, and probably the entire world like Eve and Pandora.

One brave journalist, however, stood up for Gillard. The *Age's* Katharine Murphy wrote that the speech was spontaneous, "a blow-up of pure frustration: volcanic and howling in intensity because the prelude to the explosion is a long period of not saying. What woman can't relate to that? We've all been there, not saying, broiling about the injustice of not saying."

Julia Baird agreed in the *New York Times*. "For the three years and three days that Julia Gillard was prime minister of Australia," she wrote, "we debated the fit of her jackets, the size of her bottom, the exposure of her cleavage, the cut of her hair, the tone of her voice, the legitimacy of her rule and whether she had chosen, as one member of Parliament from the opposition Liberal Party put it, to be 'deliberately barren.' The sexism was visceral and often grotesque."

Looking back, Gillard recalled, "That speech brought me the reputation of being the one who was brave enough to name sexism and misogyny. And it brought with it all the baggage that stops women naming sexism and misogyny when they see it: I was accused of playing the gender card, of playing the victim. Dumb, trite arguments that entirely miss the point: someone who acts in a sexist manner, who imposes sexist stereotypes, is playing the gender card. It is that person who is misusing gender to dismiss, to confine, to humiliate: not the woman who calls it for what it is. Calling the sexism out is not playing the victim, it is the only strategy that will enable change. What is the alternative? So the sexism is never named, never addressed, nothing ever changes?"

Several years ago, when Francesca Donner, former gender director of the *New York Times* and editor of the "In Her Words" column, worked at the *Wall Street Journal*, she wrote seven articles published on seven consecutive days exploring major issues confronting American women. Domestic abuse. Access to healthcare. Unpaid labor. When her introductory column came out, she was inundated with online criticism calling her a whiner, a complaining Karen, spoiled, ludicrous. "Why is it," she said in an interview for this book, "that when women raise an issue of inequality we are called out for whining and complaining? When women point out that they receive worse treatment from their doctors, they are told to shut up, deal with it, don't rock the boat? When women raise issues, we shove them back into their hole and tell them to get over it. Why are we not interested in hearing their stories?"

We are not interested in hearing their stories because nothing is more alarming to the Patriarchy than a fed-up woman speaking the truth about misogyny.

CHAPTER 7

WHO IS TAKING CARE OF HER HUSBAND AND CHILDREN?

In an election, if you are married, you are
neglecting him; if you are single, you couldn't
get him; if you are divorced, you couldn't keep
him. If you're widowed, you killed him.

—Barbara Mikulski, US Senator
from Maryland, 1987–2017

In a Greek myth, Atalanta, daughter of King Schoeneus of Boeotia, was a young woman of incredible athletic prowess. She could outwrestle, outhunt, and outrun all the men and had no desire to ever marry or have children. Desperate for grandchildren, her father badgered her until he obtained her promise that she would accept the man who could beat her in a footrace, with the proviso that those who lost the race would be executed. This condition was pretty effective in thinning the ranks of her suitors.

Still, many good men died in the (literal) pursuit of Atalanta. But a young man named Hippomenes obtained three golden apples from a goddess, and each time Atalanta raced ahead of him, he threw one in front of her. The third time she stopped to pick up the glittering golden bling (even huntresses couldn't resist glittering golden bling), he raced ahead of her, winning the race and her hand. Never mind that he cheated.

And so, the story has a happy ending. Atalanta stopped embarrassing all the men and found her rightful place as wife and mother. The world was once more as it should be. The Patriarchy was secure. Because women shouldn't run faster than men. Or win more honors than men. Or, heaven forbid, reject them and refuse to have children.

Fast-forward two thousand years, and another Atalanta outrunning men (figuratively) refused to marry and bear children. Elizabeth I ascended the throne in 1558 at the age of twenty-five—already long in the tooth in an era when royal girls often married at fifteen. When she showed no great interest in marrying anytime soon, the House of Commons argued that "nothing could be more repugnant to the common good than to see a Princesse . . . lead a single life, like a vestal nun." As the years passed, sometimes her council members would go down on their knees, weeping and begging her to marry.

And, indeed, there were urgent reasons for her to do so. Rulers needed direct heirs to prevent cousins—some of them possibly foreign kings—from waging civil war as they grabbed for the crown. In 1559, Elizabeth's own members of Parliament described the horrors awaiting them if she did not have an heir: "the unspeakable miseries of civil wars, the perilous intermeddlings of foreign princes, with seditions, ambitions and factious subjects at home, the waste of noble houses, the slaughter of people, subversion of towns, unsurety of all men's possessions, lives and estates." The orderly transition of power from one generation to the next was of paramount importance to every nation.

But Elizabeth was well aware of the disaster of her father's many marriages. Henry VIII had beheaded her mother and another wife, divorced two others, lost one in childbirth, and come close to burning the last one at the stake—certainly not a record that would encourage young Elizabeth to dream of marital bliss. By the time she took the throne, she had also witnessed her older half-sister's disastrous marriage to Philip of Spain. Thirty-eight-year-old Mary I's betrothal to the twenty-seven-year-old was so deeply unpopular that a rebellion broke out. Many Protestant Englishmen didn't want a Catholic Spanish king who might turn the country into a Spanish colony.

Under the marriage contract, Philip received equal titles and honors to those of his wife. The two of them appeared on coins in profile, facing each other, with the crown of England levitating magically above them. An Act of Parliament stated that Philip "shall aid her Highness . . . in the happy administration of her Grace's realms and dominions." In doing so, he dragged England into Spain's war with France, losing Calais, the last English possession of what had once been large medieval English territories in France. Mary drained both her treasury and her armories to assist

Philip in his foreign wars. If anyone had attacked England, it could not have defended itself.

Philip abandoned his devoted wife for most of the marriage, ruling his domains in Spain and the Netherlands. Mary died at forty-two, brokenhearted. Philip, who was in Brussels, wrote to his sister coldly, "I felt a reasonable regret for her death."

A medal from Mary's reign shows, on one side, the two monarchs seated on thrones side by side. Philip is on the left, the position of greater power and honor. On the other side of the medal, Mary is on the left, on a horse. Yet she is squeezed behind Philip and his horse, who are front and center. Just looking at the medal must have sent shivers of disgust through Elizabeth. She knew that any prospective husband would likely try to take her power, and even if he didn't, her counselors and nobles would naturally look to him for decisions rather than to her. Moreover, marrying a prince like Philip would bring unwelcome foreign meddling in English affairs and might drag the nation into more continental wars. But by marrying even the noblest of her subjects, she would demean herself and likely cause rival family factions at court.

The day after her accession, she appointed a thirty-nine-year-old courtier, Sir William Cecil, as her secretary of state. Cecil believed he and the council would be ruling for her until such time as they would work with her husband to govern the realm. When he discovered that an ambassadorial letter from a foreign court had been taken directly to Elizabeth without first being given to him, he was deeply concerned that the queen of England was interfering in running the country. Cecil berated the messenger, telling him he should not have brought it to her majesty, "a matter of such weight being too much for a woman's knowledge."

Similarly, when a French ambassador arrived at court, he asked for the council to attend his audience with the queen, a clear sign that he was going to discuss matters beyond the understanding of a woman. When Elizabeth heard of his request, she wrote him angrily, "The ambassador forgets himself in thinking us incapable of conceiving an answer to his message without the aid of our Council. It might be appropriate in France, where the King is young, but we are governing our realm better than the French are theirs."

In February 1559, the Commons sent the queen a formal petition at her Palace of Whitehall informing her it would be beneficial for her "and her kingdom if she would take a consort who might relieve her of these labors, which are only fit for men." If she remained "unmarried and, as it were, a vestal virgin," such a thing would be "contrary to the public respects."

Interestingly, Elizabeth fretted that any children of hers might "grow out of kind, and become perhaps ungracious," meaning that a son might grow up to take the throne from her, something that would surely gladden the hearts of her sexist counselors. She concluded that she would listen for God's direction in the matter and, though she listened attentively for many years, He never did instruct her to marry.

We cannot picture Elizabeth jammed into the side of a coin with her husband—the king, thanks to her—front and center. Or envision her perched on the less-honorable right side, while he sprawls joyously on the more prestigious left. Nor can we see her sitting sedately in the corner of the Star Chamber embroidering, while her husband—the king, thanks to her—sits at the head of the table deciding matters of state with his advisors. It is impossible to imagine her lifting her eyes from her needle now and then to cast him a radiant, approving smile for handling such weighty affairs of state that were far too much for her poor little brain.

Soon after her succession, a German envoy noted, "The Queen is of an age when she should in reason and—as is a woman's way—be eager to marry and be provided for. For that she should wish to remain a maid and never marry is inconceivable." A husband, he added, could share "the cares, the labors and fatigues of her government." But Elizabeth didn't want anyone's hands on her scepter.

In 1564, the Scottish ambassador, the perceptive Sir James Melville, told the queen, "You think if you were married, you would only be a queen of England, and now ye are king and queen both. You may not endure a commander."

Speaking to the French ambassador about a potential match with Charles IX, she said she had no intention of allowing any husband of hers to usurp control of her treasury, army, or navy. To the Spanish ambassador, she said, "There is a strong idea in the world that a woman cannot live unless she is married, or at all events that if she refrains from marriage she does so for some bad reason."

Yet how to manage the public perception that there was *something wrong with her* if she didn't marry? For one thing, other than a few stray comments, she didn't make clear her visceral fear of marriage. For twenty years, she would do a diplomatic dance with numerous foreign bachelors, keeping everyone hoping and armed invasion at bay. The first suitor was her brother-in-law Philip of

Spain—not a likely choice, considering. He soon dropped his suit of the troublesome heretic and married a properly submissive fourteen-year-old French princess.

In 1560, crazy Prince Erik of Sweden wouldn't take Elizabeth's "no" for an answer and set sail in a ship filled with costly gifts to woo her, only to be beaten back to Sweden by storms. When he tried to set out again, more storms battered his flotilla and sent him home. Before he could set out a third time, his father died, and the new King Erik XIV couldn't go a-wooing right away. Elizabeth declared the storms to be an act of God protecting her, and she was probably right. Erik would go on to stab a nobleman to death in a fit of rage, marry a tavern wench, suffer imprisonment by his brother for his foaming-at-the-mouth madness, and die of poison the jailer put in his pea soup.

Also in 1560, the twenty-three-year-old James Hamilton, Third Earl of Arran, an heir to the Scottish throne, decided to woo her and was turned down. He, too, became insane, believing murderers and witches were out to get him, and was imprisoned from 1562 until his death in 1609.

In 1568, unhappily married to his second wife, Ivan the Terrible of Russia decided he wanted to shut her up in a convent and marry Elizabeth. When she gently rebuffed him, Ivan wrote, "Thyself thou art nothing but a vulgar wench, and thou behavest like one! I give up all intercourse with thee. Moscow can do without the English peasants." As the years passed and Ivan took and executed more wives, and more subjects, and beat his son's brains out with his staff, Elizabeth must have thanked heaven for her single status. Here was a monarch who made her father look positively civilized.

While Elizabeth dodged these and other marital bullets, she still had to quieten her nobles and her people. Cleverly, she found a way to turn the argument to her advantage. She *wasn't* unmarried. She *wasn't* childless. She was "already bound unto a Husband, which is the Kingdome of England" and went on to tell her council to "reproach me no more, that I have no children: for every one of you, and as many are English, are my children." In fact, she had *loads* of children, more than three million of them.

In her last speech to Parliament in 1601, two years before her death, she again referred to herself as a mother. "And I assure you all," she said, "that though after my death you may have many stepdames, yet shall you never have a more natural mother than I mean to be unto you all." She added, "There is no jewel, be it of never so high a price, which I set before this jewel; I mean your love."

On her tombstone in Latin, she is referred to as "the mother of her country, a nursing mother to religion and all liberal sciences, . . . and excellent for princely virtues beyond her sex."

Additionally, she developed the cult of the Virgin Queen, slyly replacing that of the Virgin Mary, seen as the greatest mother of all time, who played a lesser role in Protestantism than under the Catholic Church. In her official portraits, the queen often wore white—the symbol of virginal purity—and adorned herself with more virgin symbolism: ropes of pearls, ermine, moons, sieves. In one famous portrait, she held a cornucopia, an emblem of abundance, symbol of a mother feeding her children. In another, she wore a pelican brooch. In the Renaissance, pelicans were thought to suck blood from their own bodies to feed their young.

Queen Victoria, who reigned over Great Britain from 1837 to

1901, was another mother figure to the English nation, though it was easier for her to maintain such an image than it had been for Elizabeth. Victoria took the traditional path, married a German princeling at twenty-one, and birthed nine babies. After generations of unpalatable male royals (her numerous uncles, free-spending rogues with mistresses and bastards; her grandfather, the stark raving mad King George III), many English subjects were fed up with the monarchy and wanted to put an end to it. But Victoria's respectable family life burnished the reputation of the royals. A dignified widow who wore mourning for forty years until her death, a benevolent matriarchal figure who remained above political parties, Victoria deflected the Misogynist's Handbook by abiding by the Patriarchy's most imperative rules.

While we can understand the urgency for monarchs of centuries ago to marry and have children, it is almost inexplicable that almost five hundred years later, female politicians—whose heirs will not inherit the throne or prevent a dynastic civil war—are also expected to. Anything else is often viewed as unfeminine. Suspicious. And the expectation of marriage usually applies only to women. When running for governor of Virginia in 1993, forty-six-year-old Democratic candidate Mary Sue Terry found that the media focused on her single marital status and childlessness. Former Ronald Reagan aide Oliver North opposed Terry's candidacy because, in part, the governor's mansion shouldn't be "a sterile building" but a home "where a man and his wife live, and with the laughter of their children." Yet the media blissfully ignored the single, childless state of two Virginia congressman, Terry pointed out to the *Daily Press.* "Nobody writes that about Rick Boucher, who is my age and has never been married, or about Bobby Scott, who was married only briefly," she said.

In 1993, forty-six-year-old Kim Campbell, Canada's Minister of National Defence and leader of the Progressive Conservative Party, vied to become the nation's first female prime minister. Her male opponent, Jean Chrétien, portrayed her as emotionally unstable, selfish, overly ambitious, and untrustworthy because she was twice divorced and childless. News coverage portrayed him as more likable, a traditional family guy. We can only wonder if Campbell would have won had the vote been open to the public rather than limited to her party members in Parliament.

A rising political star, Angela Merkel decided to marry her longtime partner before she became head of her party in 1998. Even that didn't help feminize her much, what with her short, no-nonsense hair, lack of makeup, and degree in quantum chemistry. During her 2005 campaign for chancellor, several women told reporters, "She's a man making it in a man's world. We don't recognize the woman in her." And yet, over her sixteen-year tenure as chancellor, Merkel became known as "Mutti" to her people, a strict and protective *Mommy*. Childless herself, she morphed into a kind of Elizabeth I mother-to-her-people figure, kindly, competent, and comforting.

During the 2010 Australian election campaign, Julia Gillard's single status was often discussed in the media. She had lived for four years with her partner, Tim Mathieson, but they had no plans to marry. Reporters couldn't understand a man living with such a powerful woman without being her husband. Some speculated that the relationship was a sham, and Mathieson must be gay. He was, after all, a hairdresser, and they had met when Gillard came in for a trim. One radio journalist, Howard Sattler, even asked Gillard if Mathieson was gay. She called his question "absurd," and Sattler was fired. Asked about it soon after, she said, "I want young girls and women to be able to feel like they can be included in public life and not have to face questioning like the questioning I faced yesterday."

New Zealand prime minister Helen Clark was persuaded to marry her long-term partner by political advisors before she ran for Parliament in 1981, despite the fact that she resolutely didn't want to marry. "As a single woman I was really hammered," she wrote in an essay in 1984. "I was accused of being a lesbian, of living in a commune, having friends who were Trotskyites and gays." She was so upset about getting married that she cried on her wedding day.

Although Clark caved to political expectations and married, she never had the children she didn't want. The press accused her of selfish ambition, putting her personal goals above those of her uterus. Many journalists questioned whether she understood the needs of families. One asked, "Is Helen Clark, childless, able to understand the concerns of parents?" Oddly, journalists never seem to ask childless male politicians whether they understand the issues of raising children.

In 2002, Clark told the *Express* that she hoped times had changed, at least in New Zealand. "I actually have great faith in the common sense of Kiwis [New Zealanders]," she said, "and I think these days most people are going to say, 'For God's sake, people are entitled to choices about their life, Helen's made her choice, that's fine with us.' So what are they getting at? Am I supposed to not be a real woman because I haven't had children? It's all bizarre and I don't think most people relate to it."

Other childless female politicians, of course, have not been so fortunate. Because not having children means you are cold, selfish, unnatural, lacking in empathy, and that there is something seriously wrong with you. And how can you make political decisions that will affect families *if you don't have children*?

Britain's second female prime minister Theresa May married her college sweetheart in 1980 when she was twenty-four, and the couple remained childless. Sometimes journalists asked outright why she had no children—as if it was any of their business—and for years she refused to answer because it was not. Perhaps she would cross her legs, swing her feet, and hope they would ask her about her leopard-print shoes instead.

In 2002, she told an interviewer pressuring her about why she had wasted her ovaries, "I don't think it's an issue. And I don't think it should become an issue." A few months later, she finally let it be known, "It wasn't a choice," adding, "But I'm not going to talk about it further." Then, in 2012, while May was serving as home secretary, she indicated in an interview with the *Telegraph* that she and her husband had wanted children but were not able to have them. "This isn't something I generally go into, but things just turned out as they did," she said. "You look at families all the time and you see there is something there that you don't have."

In 2016, she spoke on the subject again, indicating her childlessness had been a great disappointment. "Of course, we were both affected by it," she told the *Mail on Sunday*. "You see friends who now have grown-up children, but you accept the hand that life deals you."

May's childlessness has dogged her time and again throughout her political career. In 2004, *Sun* columnist Jane Moore wrote, "In his shadow cabinet reshuffle, Michael Howard has appointed Theresa May as 'spokesman for the family.' Mrs. May has no children. Politicians never learn, do they?"

"I used to be shadow Transport Secretary," May told the *Sunday Telegraph*, referring to the position in the Westminster system of government where every official cabinet member has a "shadow" duplicate in the opposition party with great visibility but no executive power. "But I'd never been a train driver. There are many types of families. A couple can be a family."

In the 2016 contest for party leadership—which would determine the next prime minister—May's competitor Andrea Leadsom told the *Times* that she differed from May because "I see myself as one, an optimist, and two, a member of a huge family, and that's important to me. My kids are a huge part of my life." She said, "I genuinely feel that being a mum you have a stake, a very real stake, a tangible stake. I have children who are going to have children who will be a part of what happens next." She tried to soften the blow by adding, "I am sure Theresa will be really sad she doesn't have children."

Mothers, she suggested, have more empathy than childless women because "you are thinking about the issues that other people have, you worry about your kids' exam results, what direction their careers are taking, what we are going to eat on Sunday."

MP Jess Phillips scoffed at Leadsom's remarks. "Are we supposed to imagine for a second that in a moment of national crisis, when Theresa May has to make a life-and-death judgment call, she is going to think, 'Ah, who cares, blow up the country. After all, it's not like I've got any kids to worry about'? I feel fairly certain that she is not on a trajectory to damn the future because she happens not to have biological offspring."

May did not comment on Leadsom's interview. But many in the press and public were furious about it. Leadsom withdrew from the running.

In retrospect, it probably redounded in May's favor that at least she had *tried* to have children. Mother Nature had, unfortunately, not obliged, but at least she wasn't so selfish as to not want any to begin with. As with Angela Merkel and Helen Clark, Julia Gillard chose to focus on her career and never wanted children. One columnist wailed that Australia was being "led by a woman who has eschewed marriage and children."

Liberal senator Bill Heffernan attacked her for not having had anything to do with diapers. "One of the great understandings in a community is family and the relationship between mums and dads and a bucket of nappies," he said. "Anyone who chooses to deliberately remain barren . . . they've got no idea what life's about." Some Australians, however, realized how ridiculous Heffernan's accusation was. When Gillard found herself stopped at a traffic light, people nearby would roll down their windows and yell, "You can borrow my kids if you like, love!"

The former leader of the Labor Party, Mark Latham, said of Gillard's perplexing decision, "Choice in Gillard's case is very, very specific. Particularly because she's on the public record saying she made a deliberate choice not to have children to further her parliamentary career. . . . I think having children is the great loving experience of any lifetime. And by definition you haven't got as much love in your life if you make that particular choice."

Not much love in her life.

The *Australian* covered the debate over Gillard's unproductive uterus under headlines such as "Barren Behavior," which compared her to a barren cow that gets slaughtered for being useless, its flesh turned to hamburger. David Farley, CEO of the Australian Agricultural Company, said, "So, the old cows that become non-productive, instead of making a decision to either let her die in the paddock or put her in the truck, this gives us a chance to take non-productive animals off and put them through the processing system. . . . So it's designed for non-productive old cows. Julia Gillard's got to watch out."

Needless to say, childless men such as President Emmanuel Macron of France are not called "barren," an insulting word replete with images of the red dusty surface of Mars. Nor are they threatened with slaughter as a useless piece of meat in an abattoir. One journalist covering the fracas over Gillard's childlessness asked, "Would the same logic apply to male politicians with low sperm counts?"

Gillard wrote in her memoir, "In being seen to offend against female stereotypes, is there anything bigger than not becoming a mother by choice? . . . It is assumed a man with children brings to politics the perspective of a family man, but it is never suggested that he should be disqualified from the rigors of a political life because he has caring responsibilities. This definitely does not work the same way for women. Even before becoming prime minister, I had observed that if you are a woman politician, it is impossible to win on the question of family. If you do not have children then you are characterized as out of touch with 'mainstream lives.' If you do have children then, heavens, who is looking after them?"

Perhaps Gillard's most ridiculous scandal occurred in 2005 when, as manager for opposition business in the House of Representatives, she cut short an international trip to deal with a political crisis back home. She hadn't even unpacked her suitcase when a photographer from the *Sunday Age* arrived to take pictures of her seated in her kitchen to accompany an article about her chances of becoming the new Labor leader. The kitchen was spartan—Gillard hates clutter—and there was no food on the counters as she hadn't had time to shop since her return. A decorative blue glass bowl sat on the table next to her.

Furor erupted when the photograph was published. Why was her kitchen so sparse? More importantly, why was her fruit bowl *empty*? The media described her kitchen as "lifeless," "eerily stark," and "unnaturally spotless." The empty fruit bowl seemed to symbolize an empty womb. An empty life. An icy, selfish heart. It is hard to imagine such howls of moral outrage if a man had allowed himself to be photographed in a bare kitchen next to an empty fruit bowl.

When Gillard became prime minister five years after the fruit bowl furor, the *Australian* reported, "She has showcased a bare home and an empty kitchen as badges of honor and commitment to her career. She has never had to make room for the frustrating demands and magnificent responsibilities of caring for little babies, picking up sick children from school, raising teenagers."

"It never occurred to me," Gillard wrote, ". . . that anyone could really contend that my life, my thoughts, my character, and my worth could be defined by the state of my kitchen. . . . No fruit. I think that was the principal flaw. No fruit. I now always have rotting bananas in the bowl just in case."

Looking back on her life in politics, Gillard wrote, "For all of our history a prime minister has been a man in a suit who has been married (to a woman) and who has children. If our first female leader also happens to be our first unmarried, childless, living with a partner, not to mention atheist, prime minister then perhaps it is not surprising that the population is having some trouble getting their heads around this new reality."

In the 1980s, as Benazir Bhutto entered her thirties, she felt more than ordinary political pressure to marry and have children due to the expectations of her traditional Muslim culture. The brilliant, Oxford-educated daughter of the murdered former prime minister Zulfikar Ali Bhutto and a rising political star, whenever she gave an interview, the journalist asked her why she hadn't yet married, a question which seemed to make her want to bang her head against a brick wall.

Many of her supporters saw her as a kind of saint, a mother to her country, a Muslim Elizabeth I who did not need to follow the normal rules. She feared alienating these people if she bowed to tradition and married. Still, in her culture, it was difficult as a single woman to have work relationships with men, particularly to socialize with them at political events. "In a Muslim society, it's not done for women and men to meet each other," she wrote in her autobiography, "so it's very difficult to get to know each other and, my being the leader of the largest opposition party in Pakistan, it would have been a lot of rumor to the grist and bad for the image if I had chosen another course."

Finally, in 1987 at the age of thirty-four, she accepted a marriage arranged by her aunt with Asif Ali Zardari, scion of a powerful and wealthy family. In announcing the nuptials, she released a statement: "Conscious of my religious obligations and duty to my family, I am pleased to proceed with the marriage proposal accepted by my mother." Were her humility and obedience simply in line with cultural and religious expectations, or do we detect a certain lack of excitement there?

She seemed to approach the wedding as if it were something like an execution. The night before the big event, Bhutto said that if she weren't in politics, "I know I would never have taken this step. I would never have gotten married at any stage." Her marriage, however, probably did help propel her into the prime minister's office the following year.

Once Bhutto married, journalists kept asking if she was pregnant (she would have three children in five years). On one occasion, Bhutto retorted, "I am not pregnant. I am fat. And, as the prime minister, it's my right to be fat if I want to."

Still, Islamists excoriated her, married or not, for being in the public eye and working closely with men who were not her relatives. And she was right to be wary of the marriage. Her husband was to become her Achilles' heel. Known as "Mr. Ten Percent" for the bribes he allegedly collected for government contracts, his corruption trashed her reputation. He was largely blamed for the collapse of both Bhutto's 1990 and 1996 governments.

During her second ouster, Zardari was thrown in jail after trying to flee to Dubai. Bhutto, too, was charged with numerous counts of corruption, though in her case it is difficult to say whether the charges were based on fact or were a purely political maneuver to keep her out of office. The two were largely living apart when she was assassinated in December 2007, though he was elected president of Pakistan nine months later in a wave of grief and support for his late wife. When Zardari published information about his private fortune upon becoming president, he revealed that it was around $1.8 billion.

WHO IS TAKING CARE OF THE CHILDREN?

For those female politicians who have done the respectable thing by marrying and having children, there is the question of why they are not home looking after them.

In her autobiography, Pat Schroeder wrote that in 1972, when the thirty-two-year-old was campaigning for a congressional seat in Colorado, "It seemed like all I was asked about was what was going to happen to my family, who would do the laundry if I was elected." When she took up her duties in the US House of Representatives, a male colleague asked her how she could be both a mother of two young children and a member of Congress at the same time. Without missing a beat, she snapped, "I have a brain and a uterus, and I use them both."

Just hours after her election as prime minister of New Zealand in 2017, thirty-seven-year-old Jacinda Ardern was asked on national television whether she planned to have children. When she had her daughter the following year, the media speculated about how she could balance family life with her political responsibilities. People were evidently worried that Ardern would be so focused on running the country that she would forget to feed her daughter, who would starve to death in her crib. Or that she would be too busy breastfeeding and changing diapers to notice when the country was ravaged by economic catastrophe, plague, wildfires, and cyclones.

Ardern's predecessor, Bill English, had first run for prime minister in 2002, when he had six children under the age of sixteen, and no one asked if he would let the country go to hell in a handbasket due to family obligations. Men rarely are. In recent years, three British prime ministers have had children while in office: Tony Blair in 2000, David Cameron in 2010, and Boris Johnson in 2020. Not a single person asked who would be looking out for the children or how the fathers would juggle family life and political responsibilities.

When Jess Phillips was first running to become a member of the UK Parliament in 2015, people asked her on a daily basis, "But what about your kids?" or "What does your husband think about this?" As she recalled in her memoir, "The latter was often said in a slightly accusatory manner, as if I hadn't told him and the questioner was going to ring him and grass on me immediately—'Your fine fellow of a husband will hear of this, you vixen.'"

After she won the election, in her very first radio interview, she was asked, "How are you going to cope with your kids?" Phillips thought the question unusually stupid. "It was almost as if I hadn't thought about the fact that becoming an MP would mean that I had to live away from my children for three days a week," she wrote, "and only now, with the help of these very wise news broadcasters, had I realized the enormity of becoming an elected representative."

She replied, "Would you ask me that if I were a male MP with children?" She estimates that she has been asked the question hundreds of times since. To point out the sexism, she often responds with something like, "Oh, those aren't my children, I just hired them from an agency to make me look more human on leaflets." But the most honest answer she gives is this one: "I'll cope with my children exactly as I did before I was an MP, very badly."

When Alaska governor Sarah Palin became Republican presidential nominee Senator John McCain's running mate in 2008, at first Republican politicians and the press lionized her for being a mother. McCain introduced her at an Ohio campaign event as "a devoted wife and a mother of five." A *Daily News* article called her "a spunky mom."

But once the thrill died down, questions arose as to how she could be vice president and a mother at the same time. In the *St. Louis Post-Dispatch*, Kurt Greenbaum asked, "Should a mother of five children, including an infant with Down's syndrome, be running for the second highest office in the land? Are her priorities misplaced?"

Palin's special-needs child prompted CNN's John Roberts to argue, "Children with Down syndrome require an awful lot of attention. The role of Vice President, it seems to me, would take up an awful lot of her time, and it raises the issue of how much time will she have to dedicate to her newborn child?" Bill Weir of ABC's *Good Morning America* asked a similar question of a McCain spokesperson: "Adding to the brutality of a national campaign, the Palin family also has an infant with special needs. What leads you, the Senator, and the Governor to believe that one won't affect the other in the next couple of months?"

More controversy arose when the campaign announced that Palin's seventeen-year-old daughter was pregnant. As the *New York Times* stated: "With five children, including an infant with Down syndrome and, as the country learned Monday, a pregnant 17-year-old, Ms. Palin has set off a fierce argument among women about whether there are enough hours in the day for her to take on the vice presidency, and whether she is right to try."

Oddly, few seemed to consider that Palin's husband might lift a finger now and then to help with his children. Sally Quinn of the *Washington Post* did, but then quickly rejected it, stating, "Everyone knows that women and men are different and that moms and dads are different and that women—the burden of child care almost always falls on the woman . . . when you have five children, one a 4-month-old Down syndrome baby, and a daughter who is 17 . . . and who is going to need her mother very much in the next few months and years with her own baby coming, I don't see how you cannot make your family your first priority."

A *Daily Planet* editorial accused Palin of selfishness by pursuing her political career even after her seventeen-year-old daughter became pregnant. She should model herself after Nancy Pelosi, the article advised, whose children were almost grown when she first ran for Congress back in 1987: "If Sarah respected the privacy of the daughter and the boyfriend, she would not have thrust herself—and them—into the spotlight at this particular difficult moment. There's no feminist ideology that mandates exploiting and neglecting your kids in order to get ahead. Nancy Pelosi, another mother of five, did it right, and Palin could, too, if she had an ounce of compassion or a grain of common sense. In other words, wait until your children are grown before pursuing such a high-profile career."

Another charming presidential candidate with young children, Senator Barack Obama, was never subjected to the same questions regarding the appropriateness of running for office with small children. It was assumed his wife would care for the children.

Over the years, Ursula von der Leyen, president of the European Commission, became fed up with being repeatedly asked how she was able to balance her career with raising her children. On a podcast for International Women's Day, March 8, 2021, she recalled the time when, as Germany's minister for family affairs, a talk show moderator asked the mother of seven, "Have you already chosen whether you want to be a bad mother or a bad minister?"

On October 14, 2020, US Supreme Court justice nominee Amy Coney Barrett underwent questioning in her confirmation hearings. Senator John Kennedy of Louisiana asked "a sincere question," as he put it. "Who does the laundry in your house?"

Her views on abortion, her stance on gay rights, the charismatic Christian group where she serves as a "handmaid"—these were all issues deserving of sincere questions. But, seriously, her *laundry?* How was that relevant? Would any male Supreme Court nominee—Antonin Scalia, for instance, who had nine children— ever be asked that question?

And let's examine, for a moment, her possible answers. If she had said, "I have a housekeeper who does it," she would have sounded elitist, out of touch with regular working Americans. If she had said she didn't get around to doing it often, she would have been painted as a bad wife, a bad mother, a selfish careerist, and a filthy housekeeper who probably had cockroaches gleefully scurrying over her kitchen counters. If she had said she did the laundry frequently with such a large household, people would have wondered how she would find the time to be a Supreme Court judge, what with her folding so many pairs of underpants.

As it was, Barrett laughed pleasantly at Kennedy's startlingly sexist question and said she and her husband were trying to get the children to do their own laundry, though those efforts were not always successful. It was the perfect answer. She came off looking like a *good mother.*

When Hillary Clinton first ran for president in 2008, her daughter Chelsea was twenty-eight, so no one could criticize Clinton for neglecting a young child. But six years later, when Clinton was looking at the next presidential election, Chelsea was pregnant with *her* first child. *USA Today* speculated, "It's unclear how Chelsea's pregnancy will affect Hillary Clinton, who is considering a race for president in 2016."

In the 2012 race, no one asked whether Republican presidential nominee Mitt Romney should stay home to help with his eighteen grandchildren, including newborn twins. Nor, in 2016, did journalists ask whether Donald Trump shouldn't focus on helping his daughter Ivanka with her newborn son.

It's only female politicians who should give up their careers to take care of children. And grandchildren. Maybe even great-grandchildren. Clearly, it would be best for mankind if they never worked at all and just stood hopefully by with a closetful of formula and diapers for any genetic progeny that might appear.

CHAPTER 8

SHE'S A WITCH AND OTHER MONSTERS

*Why is it when a woman is confident and
powerful, they call her a witch?*

—Lisa Simpson

G oing back thousands of years, powerful women have been linked to bubbling cauldrons, spellbooks, and eye of newt in a glass jar. Diminishing women by calling them witches is probably the oldest page in the Misogynist's Handbook. And sure enough, no sooner had Kamala Harris been named as Joe Biden's running mate than a GIF appeared of her as the Wicked Witch of the West, with a green face and pointed black hat. Harris's sudden metamorphosis into a witch was, in fact, more a badge of honor than an insult, proof that a woman is getting under the skin of those who don't want her in power.

On April 7, 2021, Newsmax journalist Grant Stinchfield, who

obviously couldn't come up with any valid criticisms of Harris, edited together several clips of her laughing, one after the other. Without seeing the part where something funny was said (such as the one where Rachel Maddow asked, "Did you see the fly on Mike Pence's head during the debate?"), Harris did indeed look deranged. On the right side of the screen was a photo of Harris with green skin and a pointed hat and the headline "Can We Talk about Kamala's Cackle?" Stinchfield played video of the Wicked Witch of the West from *The Wizard of Oz*, the three witches from the film *Hocus Pocus*, and the cartoon evil witch queen from *Snow White and the Seven Dwarfs* proffering the poisoned apple. Then he said, "Nancy Pelosi, move over. There's a new witch in town, and her name, of course, is Cackling Kamala. Oh, how sad is that."

Margaret Thatcher, tough, opinionated, and loudly critical, was often called a witch. In 1971, as secretary of state for education and science, she cut subsidies that gave free milk to elementary school children (most of it went unused as many kids didn't want milk—they wanted soda). But she was caricatured in the press as a broomstick-riding wicked witch snatching milk from children. As prime minister, her take-no-prisoners stance on union busting and budget cutting resulted in unpopularity in some circles, fueling the witch comparisons. When she died in 2013 at the age of eighty-seven, "Ding-Dong! The Witch Is Dead" became a top hit in Britain, seventy-four years after it first appeared in the soundtrack to *The Wizard of Oz*.

As prime minister of Australia, Julia Gillard often had to push through crowds of protesters carrying signs that read "Ditch the Witch!" Nancy Pelosi was portrayed as a witch on a broomstick on T-shirts with the text "This is my Nancy Pelosi costume." And when she was UK prime minister, Theresa May was filmed laughing loudly, a "witch's cackle" that quickly went viral.

Hillary Clinton has been called a witch more than any modern figure. When she was first lady, some of her Secret Service officers dubbed her airplane "Broomstick One." A CNN commentator called her "the Wicked Witch of the West." Posters and T-shirts of a green-faced Clinton on a broomstick, wearing a pointed black hat, abounded at campaign rallies of her political opponents in 2008 and 2016. In 2018, a San Diego resident fashioned drones of Clinton riding a broomstick and Trump flying all by himself like Superman and flew them over the city to the alarm of many residents.

The sexist trope of witch is closely related to those of other monsters and devils. In 2014, Montana Republican congressional candidate Ryan Zinke called Clinton the "anti-Christ" at a January campaign event. At the Republican National Convention that summer, Ben Carson linked her to Lucifer himself. Referring to Clinton's appearance at the 2016 Democratic National Convention, far-right radio show host Alex Jones said, "She's a creep, she's a witch, she's turned over to evil. Look at her face. . . . All she needs is green skin." In October of that year, Jones claimed she reeked of sulfur. Literally demonizing her, anti-Hillary groups generated memes of her with digitally simulated devil horns and the number 666 tattooed across her forehead. Even liberal and former MSNBC commentator Chris Matthews called Hillary Clinton "witchy" and a "she-devil." In 2016, Bernie Sanders's supporters created the hashtag #BERNTHEWITCH.

What is at the root of calling women witches? Probably the same thing as the cause of misogyny itself. Magic is a sign of great, dangerous power that must be carefully controlled. The magic of bringing forth human life has always been a woman's sole prerogative, the connection of her body to the phases of the moon a dark and troubling mystery. And the powerful sexual desires women arouse in men render them vulnerable, not in control, and hating the cause of those desires that weaken and distract. To transform such physically and intellectually superior beings as men into lustful fools, women must be using dark magic.

The biblical Book of Exodus tells us, "Do not allow a sorceress to live" but doesn't mention anything about sorcerers. Historians estimate that in the witch hunts of the fifteenth to eighteenth centuries—during which somewhere between 60,000 and millions of innocent people were burned alive or hanged—eighty percent of those executed were women. The definitive treatise on witchcraft, the 1486 *Malleus Maleficarum*, or the *Hammer of Witches*, informs us, "Three general vices appear to have special dominion over wicked women, namely, infidelity, ambition, and lust. Therefore, they are more than others included towards witchcraft, who more than others are given to these vices." As the inquisitors explained, "All witchcraft comes from carnal lust, which is in women insatiable." (And men are, as we all know, generally uninterested in sex.)

The earliest female sex spirit we know of, Lilith, has wandered the earth for four thousand years on demonic wings, first among the ancient Babylonians, Hittites, Egyptians, Israelites, and Greeks, where she caused pregnant women and infants to sicken and die. Lilith flew briefly into the Bible, where the prophet

Isaiah avoided her in the wilderness. With her taloned feet and perky breasts filled with poison instead of milk, the sexually insatiable demon-woman ravished men as they slept, causing them to produce nocturnal emissions, from which she became pregnant, breeding more demons.

The most horrifying characters of ancient Greek mythology were the Furies, a trio of female spirits of vengeance who hounded and whipped those who broke oaths or mistreated the aged and their parents. Far older than the Olympian pantheon, the Furies sprang from the drops of blood resulting from the castration of Uranus by his son, the Titan Cronus. Often portrayed as bat-winged crones with snakes for hair, their names meant "endless anger," "jealous rage," and "vengeful destruction." These daughters of castration symbolized the horrors that could unfold if men lost their virile power and women took over.

Their mythological cousins, harpies, were hunger-crazed monsters with a woman's face and breasts, bird's wings, and bird's feet with sharp talons for shredding prey. Known for their foul smell, they abducted and tortured souls on their way down to Hades. Their name has come down to us today to describe a nagging or shrewish woman.

Just as there were no male Furies or harpies, there is no masculine word that exactly equates to witch. We don't picture a warlock with warts, green skin, and chin hairs, wearing a pointed hat and flying around on a broom. We don't see him stirring a bubbling cauldron in *Macbeth* and eating children in *Grimms' Fairy Tales*. "Wizard" brings up images of tall, stately, powerful men with shining white beards, like Gandalf or Dumbledore.

In addition to the green-faced, haggle-toothed witch, there is another kind: the cold-hearted, manipulative beauty with prominent cheekbones à la Maleficent and the Evil Queen in *Snow White and the Seven Dwarfs* who, incidentally, instead of ruling her kingdom wisely, spent all her time obsessing about her appearance. These archetypes hark back to the ancient world, the myth of woman as sorceress and enchantress. Circe, a character in Homer's *Odyssey*, turned men who washed up on her island shores into swine. The sirens' heartbreakingly beautiful song lured sailors upon the rocks.

The maenads—whose name means "the raving ones"—were mortal women imbued with dark power by the god Dionysus, whom they worshipped. The most dramatic example of what could happen to out-of-control, witchy women, the maenads left their homes—and husbands—to live free in the wild, where they wore fawn skins, carried large sticks called *thyrsi* wrapped with ivy, and draped themselves in living snakes as jewelry. In their wildest drunken frenzies, they tore men and animals to pieces and ate them raw.

The German writer Walter Friedrich Otto, an expert on ancient Greek myths, wrote, "They strike rocks with the thyrsus, and water gushes forth. They lower the thyrsus to the earth, and a spring of wine bubbles up. If they want milk, they scratch up the ground with their fingers and draw up the milky fluid. Honey trickles down from the thyrsus made of the wood of the ivy, they gird themselves with snakes and give suck to fawns and wolf cubs as if they were infants at the breast. Fire does not burn them. No weapon of iron can wound them, and the snakes harmlessly lick up the sweat from their heated cheeks. Fierce bulls fall to the ground, victims to numberless, tearing female hands, and sturdy trees are torn up by the roots with their combined efforts." Most horrifying of all, according to the Patriarchy, as these women ran wild their husbands were sitting home with no dinner and no sex.

Princess Medea of Colchis, Circe's niece, used spells and potions to help the hero Jason find the golden fleece; she pulled the moon from the sky, called the dead from their graves, and made rivers flow backward. Medea's frightening connection to the beating heart of all nature is clearly stated in this passage from *Bulfinch's Mythology*. When her lover Jason wanted her to add years to the life of his aged and ailing father, "The next full moon she issued forth alone, while all creatures slept. Not a breath stirred to foliage, and all was still. To the stars she addressed her incantations, to the moon, to Hecate, goddess of the underworld, and to Tellus goddess of the earth, by whose power plants potent for enchantments are produced. She invoked the gods of the woods and caverns, of mountains and valleys, of lakes and rivers, of winds and vapors. While she spoke, the stars shone brighter and presently a chariot descended through the air, drawn by flying serpents. She ascended it and borne aloft made her way to distant regions, where potent plants grew which she knew how to select for her purposes." She did indeed restore Jason's father to youth and health.

But witchcraft cuts both ways. When the faithless Jason wished to marry a young virgin, Creusa, princess of Corinth, and put away Medea, the sorceress sent a poisoned robe as a gift to the bride, killed her own children, set fire to the palace, mounted her serpent-drawn chariot, and flew away.

In ancient Roman lore, witches were gray-haired hags who murdered and tortured children, even those still in the womb, to harvest body parts to use in their spells. They were also thought to devour children alive. Their tamer activities included making love potions and casting curses. Men believed that a witch's most disturbing spell, however, caused sexual impotence. The first-century CE Roman poet Ovid blamed his inability to get an erection on a witch who cast a spell on a kind of voodoo doll to interfere with his lovemaking. (It was a popular excuse that must have caused many a disappointed woman to roll her eyes.)

Fifteen hundred years later, in 1453, King Enrique the Impotent of Castile divorced his wife after thirteen years—she was still a virgin—claiming that an unknown witch must have put a spell on his penis. He took a second bride—a ravishing sixteen-year-old—in the hopes that the witch's spell would not work with this one. On the wedding night, he fortified himself with the Viagra of the time: a broth of bulls' testicles mixed with powder of porcupine quills. Unfortunately, the witch's spell still proved effective.

It's probably no coincidence that the *Malleus Maleficarum* accused witches of stealing penises. "Finally, what shall we think about those witches who somehow take members in large numbers—twenty or thirty—and shut them up together in a birds' nest or some box, where they move about like living members, eating oats and corn? This has been seen by many and is a matter of common talk. . . . A man reported that he had lost his member and approached a certain witch in order to restore his health. She told the sick man to climb a particular tree where there was a nest containing many members and allowed him to take any one he liked. When he tried to take a big one, the witch said you may not take that one, adding, because it belonged to a parish priest."

Tucker Carlson, when hosting a talk show on MSNBC, frequently described Clinton in emasculating terms, such as, "There's just something about her that feels castrating, overbearing, and scary," adding, "When she comes on television, I involuntarily cross my legs." On another occasion, he called Hillary Clinton the "anti-penis" and said, "You look at Hillary and you know in your heart that if she could castrate you, she would." MSNBC's Chris Matthews dubbed Clinton's male supporters "castratos in the eunuch choir."

In 2008, a novelties company channeled this powerful male fear of impotence and castration into the manufacture of Hillary Clinton nutcrackers. The device was a pantsuit-clad Clinton doll who opens her legs to reveal stainless steel thighs that literally busted nuts.

"SHE BEGUILED MANY PEOPLE THROUGH HER SATANIC WILES"

In the early fifth century, a time when the increasingly powerful Catholic Church clashed with the remainders of the pagan world, Hypatia of Alexandria was a renowned pagan mathematician, scientist, teacher, and philosopher. Known for both her genius and her acceptance of people of all religions, she gained great power as top politicians routinely asked her advice, particularly on ethics.

But there were many public officials who did not want a woman to have that kind of influence, especially one questioning their ethics. When the Roman imperial prefect Orestes turned to her regularly for advice on how to handle the unruly population, her political opponents said she had enchanted him with witchcraft, being "devoted at all times to magic, astrolabes and instruments of music, and she beguiled many people through her Satanic wiles." Her enemies, evidently at a loss at what to do with scientific instruments, assumed she was up to no good with them.

As conflict increased between Orestes and Bishop Cyril of Alexandria—no great fan of Hypatia's—Cyril's followers blamed the witch, now a venerable sixty-five years old, for preventing peace and prosperity in the city. In March 415 CE, a mob seized her from her carriage, dragged her into a church, carved out her eyes and her living flesh with oyster shells, then tore her limb from limb. As if that weren't enough already, they then carried the pieces of Hypatia outside the church and set them on fire.

The dazzling Anne Boleyn was also called a witch, and many at the time believed she had used witchcraft to become queen. Eustace Chapuys, the Spanish ambassador to England, wrote that the king "had been seduced and forced into this second marriage by means of sortileges and charms." Because, as we all know, Henry VIII was a weak, vacillating sort easily overcome by a woman whispering incantations over a candle.

Though Anne's hair was likely auburn, she was posthumously given black hair to make her seem more witchlike. In Anne's lifetime, the Venetian ambassador described her hair as "*marrone*," which can refer to a range of shades from brown to auburn. Her daughter Elizabeth I wore a ring—now called the Chequers ring, named after the prime minister's country residence, where it resides in an antiquities collection—that opened up to show two miniatures, one indisputably of herself, the other, most experts agree, of her mother. No other Tudor court personage wearing a 1530s headdress could conceivably be the person whose image Elizabeth, born in 1533, would have wanted in her ring. Yet the Anne figure has red hair.

Another clue to Anne's identity is that the miniature was hidden, covered by rubies, diamonds, a pearl, and a bit of blue enamel in the shape of the letters *ER* (Elizabeth Regina). The ring had to be opened to see her. Elizabeth rarely, if ever, mentioned her mother. In 1536, Henry VIII had Anne Boleyn executed on trumped-up charges of adultery, annulled their marriage, and officially declared Elizabeth a bastard. Catholic kings, considering her to be an illegitimate queen, were itching to topple her and place a Catholic rival on the throne. Any mention of Anne Boleyn would just fan the flames of the troubling question regarding Elizabeth's right to rule. Because of this, she never had her mother's bones dug up from the floor of the Tower church, where they moldered in an arrow box, for a more dignified burial in Westminster Abbey.

How, we ask, did auburn-haired Anne Boleyn end up in the popular imagination with black hair? Well, for one thing, no known portraits of her survive from her lifetime, probably because Henry VIII had them destroyed. (It's so irritating to be reminded that you beheaded your wife.) In 1585, a Catholic propagandist named Nicholas Sander was the first to mention black hair—everybody knows witches have black hair—as well as a sixth finger and several disfiguring moles, all signs of a witch. Sander, who hoped Elizabeth would indeed be toppled, even gave Anne a third nipple, the devil's teat. It is difficult to believe that suspicious, superstitious Henry, a man with a keen eye for female beauty, would be attracted to a woman with all the marks of a witch.

True, Anne was ambitious—even more disturbing then than now—and she had a fiery temper. But it was her power that unsettled and enraged, power no woman should have, particularly one not born into a royal family. Henry consulted Anne on both political and religious policy, creating conflict as many Englishmen welcomed the new religion while many others clung to the old. The king's next queen, silent Jane Seymour, with the seal of Patriarchal approval upon her, inspired no outrage as she possessed no power, nor is it likely she would have known what to do with it if she had. No one ever accused Jane of using witchcraft to become queen, even though Henry married her a mere eleven days after beheading her predecessor whose satanic powers had somehow failed to prevent her own execution.

Catherine de Medici was another woman who, many men believed, usurped power that should have belonged to them and probably did it through witchcraft. True, she consulted astrologers, but so did every other monarch of the time, though hers seem to have made more accurate predictions than most. Their success—which may have been embellished a bit over time—just reinforced her reputation as a necromancer. For instance, in 1556, when she asked the astrologer Nostradamus to cast the horoscopes of her family, he reportedly went into a trance and wrote in his book of prophecies:

The young lion will overcome the older one,
on the field of combat in a single battle;
He will pierce his eyes through a golden cage,
Two wounds made one, then he dies a cruel death.

The astrologer advised Catherine's husband, Henri II, not to joust, as he would suffer a horrible accident, a prediction the king blew off. Three years later, the lance of his younger opponent crashed through the king's golden helmet (the golden cage) and splintered into two pieces, one of which impaled his eye, the other his temple (two wounds made one). Some spectators claimed the men carried shields with lions on them, a common heraldic theme. The king died after ten days of absolute agony.

From the day of Henri's death, Catherine wore only black mourning, which also contributed to her reputation as a witch. (Everyone knows witches wear black.) Her Italian astrologer Cosimo Ruggeri is reputed to have correctly predicted that three of her sons would become king, and none of them would have legitimate sons. In 1574, the imperial ambassador wrote Philip II of Spain that the queen and Ruggeri had conducted a black mass at a black altar with black candles, where they sacrificed a young Jewish boy and cut off his head, hoping to hear it speak occult secrets.

Catherine was accused of murder in an immensely popular 1575 book—the title is quite a mouthful—*A Mervaylous Discourse upon the Lyfe, Deeds, and Behaviours of Katherine de Medicis, Queen*

Mother: wherein are displayed the meanes which she had practised to attain unto the usurping of the Kingedome of France, and to the bringing of the estate of the same unto utter ruine and destruction. According to this book, she tried to murder an enemy with a poisoned apple, just as the evil witch queen did to poor Snow White. The anonymous author even compared her to the mythical Circe, for "with her ensorcelled drinks she had bewitched us and transformed us into the shapes of bruit beasts or rather deprived us of our senses."

The underground French press portrayed Marie Antoinette as a winged harpy whose razor-sharp talons dripped blood from her victims, Satan's daughter who drank and bathed in human blood, and an insatiable monster out for human flesh. Her head was printed on a four-legged animal's body, with snakes writhing in her hair.

Such antiquated references to powerful women as monsters are alive and well in the twenty-first century. The London *Sunday Times* called Hillary Clinton an "unkillable" zombie moving "relentlessly forward." A commentator on Fox News called her a blood-sucking "vampire." The day after Kamala Harris's successful vice presidential debate against Mike Pence, Donald Trump called her "this monster." How odd that we have made absolutely no progress since John Knox wrote in his 1558 *First Blast of the Trumpet Against the Monstrous Regiment of Women* that any woman who dared "to sit in the seat of God, that is, to teach, to judge, or to reign above a man" was "a monster in nature."

Transforming powerful women into green-faced witches and snake-haired monsters serves a dual purpose: it simultaneously diminishes the power of those who have escaped the patriarchal boundaries on their place and also discourages other women from following in their path. Who in their right mind would want to be chopped to pieces like Hypatia, Anne, or Marie? Who would want to be trashed in the media like Clinton, Gillard, and Harris?

Hillary Clinton has been accused of participating in ritual sex magic and attending a "witch's church" (whatever that is) with her female friends. By early 2019, right-wing religious groups were accusing socialist representative Alexandria Ocasio-Cortez of belonging to "a coven of witches that cast spells on Trump 24 hours a day" (which actually might explain a lot).

Well might we wonder what was truly going in Hillary Clinton's scary basement with her sinister private email server when she was secretary of state. Flickering torches on the walls? Naked crazed dancers smeared with blood? Female Democrats, dressed in fawn skins like maenads, waving ivy-bound sticks and muttering incantations? Stolen penises of Republican politicians and Tucker Carlson, squirming around in a box, eating oats and corn?

CHAPTER 9

SHE'S A BITCH AND OTHER ANIMALS

When a man gives his opinion, he's a man. When a woman gives her opinion, she's a bitch.

—Bette Davis

On July 20, 2020, Representative Ted Yoho (R-FL) called the wrong woman "bitch." Representative Alexandria Ocasio-Cortez of New York was on her way up the Capitol steps to cast a vote when Yoho accosted her, furious that she had recently associated a spike in crime with poverty. "I was minding my own business walking up the steps," she explained in her speech on the House floor three days later, "and Representative Yoho put his finger in my face. He called me disgusting. He called me crazy. He called me out of my mind. And he called me dangerous." The congresswoman told him he was rude and kept going.

Yoho then called her a "fucking bitch," according to reporters who witnessed the exchange. Bear in mind, the word "bitch" is the blazingly angry form of "unlikable," manifesting not only misogyny but raging misogyny. And "fucking bitch" is raging misogyny on steroids.

On July 22, Yoho issued a bizarre, sexist fauxpology. After explaining that he hadn't meant for Ocasio-Cortez to hear his comment—and, in fact, she had not—he invoked his wife and daughters as a kind of giant silver crucifix hoisted against the forces of darkness produced by his own misogyny, and plowed ahead with a wildly unconnected comment, "I cannot apologize for my passion or for loving my God, my family, and my country." It remains unclear why his being in possession of a wife and daughters, as well as his love of God, family, and country, would cause him to verbally attack a fellow member of Congress, but he probably felt the bitch made him do it. He said he hadn't actually said those words *to her*, but if those words he didn't say to her "were construed that way I apologize for their misunderstanding." (Though, honestly, it is a bit hard to understand how the words "fucking bitch" could be misconstrued, even when lobbing them into empty air Ocasio-Cortez had just vacated.)

The following day, Ocasio-Cortez uttered an epic ten-minute takedown that will live as one of the most brilliant feminist speeches of all time alongside the Misogyny Speech of Julia Gillard. "This harm that Mr. Yoho tried to levy at me was not just directed at me," she said calmly. "When you do that to any woman, what Mr. Yoho did was give permission to other men to do that to his daughters. . . . I am here to say, that is not acceptable. . . . Having a daughter does not make a man decent. Having a wife does not make a decent man.

Treating people with dignity and respect makes a decent man. And when a decent man messes up, as we all are bound to do, he tries his best and does apologize," she said.

"I am someone's daughter, too. . . . And I am here because I have to show my parents that I am their daughter and that they did not raise me to accept abuse from men. . . . You can be a powerful man and accost women. You can have daughters and accost women, without remorse. You can be married and accost women. You can take photos and project an image to the world of being a family man, and accost women, without remorse, and with a sense of impunity. It happens every day in this country."

Speaking to reporters later that day, House Minority Leader Kevin McCarthy of California said he thought Yoho's response was sufficient. "In America, I know people make mistakes," he said. "We're a forgiving nation. I also think when someone apologizes, they should be forgiven. I don't understand that we're going to take another hour on the floor to debate whether the apology was good enough or not."

Because boys will be boys and locker room talk.

Ocasio-Cortez's put-down for the ages struck a nerve with millions of women. We have all been called bitches, even fucking bitches. And there is no similar epithet to levy at men. Bitch is a word pertaining solely to the female gender. Bastard, son of a bitch, prick, asshole—they are pale, weak, lame nouns, lacking that perfect vicious zing, the twist of the verbal stiletto. (It's just like *witch*—zap! And *warlock*—meh.) Even the sound of the word *bitch* is reminiscent of a slap.

The use of *bitch* to refer to women started in the fifteenth century, right around the same time as witch hunting, oddly enough. Bitches were hunting dogs that needed to be disciplined and controlled. In the 1760s, the renowned misogynist King Frederick the Great of Prussia had three dogs—bitches all—that he named after Europe's three most powerful women: Empress Catherine the Great of Russia, French royal mistress Madame de Pompadour, and Austrian empress Maria Theresa. He was delighted that when he snapped his fingers, the bitches came running.

Marie Antoinette, too, was called a bitch. But in her case, the underground press made a pun of the French word for an Austrian woman: *l'autrichienne*. The French word for female dog, or bitch, is *chienne*. Her nickname, the Austrian bitch, was used in countless pamphlets to belittle and scorn the glamorous foreign woman—the talk of Europe—who refused to sit silently in the background as other recent French queens had.

Kory Stamper, lexicographer and author of *Word by Word: The Secret Life of Dictionaries*, told *HuffPost*, "Calling a woman a bitch tells her that she's too loud, too forward, too obnoxious, too independent, too-too. Calling her a bitch reminds her that she should, like a hunting dog, be controllable."

Interestingly, the use of the word *bitch* more than doubled between 1915 and 1930, a time when women in the US and many other nations received the right to vote, according to a 2014 *Vice* story by Arielle Pardes. Clearly, the more agency women have, the bitchier they become.

They have never been quite so bitchy as in recent years, daring, as some have, to run for the highest elected office in the land. Harlan Hill, whose online bio identified him as a member of the

Trump campaign advisory board, tweeted in October 2020 a statement so offensive and misogynistic about Kamala Harris that even Fox News decided it was too much and he would no longer be welcome on the network. Without a shred of remorse, Hill told *Mediaite*: "I stand by the statement that she's an insufferable power-hungry smug bitch."

At Donald Trump's 2016 rallies, T-shirts and hats encouraged voters to "Trump That Bitch," a clear reference to Hillary Clinton, or offered the motto: "Life's a bitch, don't vote for one." Conservative pundit Ted Nugent has called Clinton a "lying America-destroying criminal ass bitch." He shared a video depicting Clinton being shot, in which he remarked, "I got your gun control right here, bitch!"

When January 6 rioter Richard "Bigo" Barnett sat at Nancy Pelosi's desk, he left a note: "Hey Nancy, Bigo was here you bitch." (Though it seems he couldn't spell "bitch," and wrote "bictch," and his handwriting was so appalling his attorney claimed he had written "you biatd"—whatever that means—as a serious part of his defense. In court.)

In 2008, when a voter asked John McCain during a campaign rally, "How do we beat the bitch?"—meaning candidate Hillary Clinton—the usually gentlemanly McCain offered no rebuke. He merely paused for a moment and then replied, "That's an excellent question."

"I believe that bitch is a metaphor that signals backlash, and backlash emerges when women are on the cusp of achieving real power in politics," Karrin Vasby Anderson, a feminist author and professor of communications studies at Colorado State University, told *Vox* the week after Harris was named as Biden's running mate. "It's a tool of containment because it's a flag of somebody transgressing a boundary."

House Speaker Nancy Pelosi, a tough leader, isn't always called a tough leader as male Speakers of the House have been. Having stepped out of the prescribed female box, having transgressed that boundary, she's often called a bitch. Amazon sold T-shirts with the image of a smiling Trump, thumbs up, and the words, "Ditch the bitch. Impeach Pelosi." Men who know what they want and go for it are called effective. Women are called bitches.

The word *bitch* "taps into and reinforces misogyny: contempt for and anger at women simply for being women," Georgetown University professor Deborah Tannen told *HuffPost* shortly after Ocasio-Cortez's speech. "Simply for being."

OTHER ANIMALS

In 2013, a scandal involving Australian prime minister Julia Gillard rocked the nation. Menugate, as it was dubbed, began when a candidate for the opposing political party created a joke menu for a fundraiser, describing the main course as "Julia Gillard Kentucky Fried Quail—Small Breasts, Huge Thighs, and a Big Red Box," the latter a reference to the red-haired prime minister's genitals. Perhaps it's not surprising that in 2016 bumper stickers appeared in the US advertising the "Hillary Meal Deal: two fat thighs, two small breasts, and a bunch of left wings."

Despite the political rancor of the Trump years, we never saw anyone chopping him up and slapping him on a menu: "Donald

Duck à la Orange, stuffed with lard, with a fat rump and tiny testicles."

Let us not forget that Gillard, who has perhaps suffered the most misogynistic treatment of any female national leader, was dismissed by the Australian Agricultural Company CEO as "a nonproductive old cow" that would be slaughtered to make hamburger meat. Many female politicians—including Kamala Harris and Hillary Clinton—are said to "cackle," like hens. Turning humans into animals or meat, a jumble of unattractive body parts to be consumed, seems to be reserved mostly, perhaps only, for women. It devalues their power, neutralizes their threat. Donald Trump called his former senior White House aide Omarosa Manigault Newman, his highest-ranking Black staffer, "that dog." He has called other women fat pigs, horse-faced, and disgusting animals. There's a reason the most common insult used for women—bitch—is an animal. Animals are less than human.

During the 1999 New Zealand election campaign, the two top contenders were women: Helen Clark and Jenny Shipley. Predictably, the media reported the contest as a "catfight." One news story said the two women "circled like wary cats during a televised party leaders' debate." Such stories reduced an election between two experienced politicians to an image of jealous women scratching each other, hissing, and pulling out hair, delegitimizing them.

Then there's the term "queen bee," a gender stereotype used to denote a woman of some authority who views other women as competition. Once again, there is no male equivalent. This phrase doesn't even call her an animal. She's an insect.

In 2013, when Prime Minister Helle Thorning-Schmidt sat

with President Barack Obama at Nelson Mandela's memorial service, *New York Post* columnist Andrea Peyser wrote, "The Danish
hellcat hiked up her skirt to expose long Scandinavian legs covered
by nothing more substantial than sheer black stockings." Referring
to the prime minister's good looks, Peyser called her a "Danish pas-
try," reducing her from a living animal (hellcat) to a lifeless mixture
of dough and marmalade.

When anyone says:

She's a bitch.

She's a pig.

She's a dog.

She's horse-faced.

She cackles.

It's a catfight.

She's a queen bee.

She is a food item.

They are stating that women are less than human. It justifies
misogyny.

CHAPTER 10

HER SEXUAL DEPRAVITY

Women who sleep around in this city are called sluts.
Men are called senators.

—Pat Schroeder, Colorado congresswoman, 1973–1997

No one is sure exactly when Mary Magdalene became a prostitute, but it was probably some five hundred years after her death. In a church actively entrenching itself in misogyny, this powerful figure in Jesus's ministry needed to be diminished and silenced.

In the four canonical gospels, Mary Magdalene traveled with Jesus and his disciples, along with other women who, according to the Gospel of Luke, Jesus had cured of evil spirits and diseases. According to Mark and Luke, Jesus had driven out "seven demons" from Mary Magdalene, which in the first century CE may have meant an illness that required seven exorcisms for complete healing. She and the other women traveled with Jesus and the disciples, "helping to support them out of their own means," according to Luke. Clearly, she was a woman of some wealth.

Mary Magdalene is mentioned twelve times in the gospels, more than most of the disciples. All four gospels state that she was present at the crucifixion and was the first—either alone or with other women—to find Jesus's empty tomb. Moreover, in Matthew, Mark, and John, she was the first person to see the resurrected Jesus. According to John, as she stood outside the tomb, Jesus appeared and instructed her to relay to his disciples a crucial message: "I am ascending to my Father and your Father, to my God and your God." In extra-canonical texts, she is referred to as one of Jesus's closest companions, and it is possible she was an early Christian leader. An early third-century Church Father, Hippolytus, calls her "apostle to the apostles" in his *Commentary on the Song of Songs*.

For three centuries after the crucifixion, church services were held in homes—the accepted domain of women. And here women played a major role—teaching, disciplining, and managing material resources. According to tombstones found in France, Turkey, Greece, Italy, and Yugoslavia, some of these women were priests. Women lost ground in the fourth century when Emperor Constantine legalized Christianity and built grand basilicas—the public sphere of men—for religious services. The Church endeavored not only to remove women from any positions of power, but also to excise any trace of their role in Jesus's ministry and the earliest beginnings of Christianity.

For instance, the apostle Junia, whom Paul hailed in Romans 16:7, was transformed into Junias, a male name that incorrectly persists in some Bibles today. The mosaic of Bishop Theodora in the ancient Roman church of Saint Prassede has had the feminine ending of her name scratched off, leaving poor Bishop Theodo wearing a woman's headdress. But as it would have been awkward turning Mary Magdalene into Marvin Magdalene, they turned her into a whore—the best way to take down a powerful woman since time immemorial.

In a series of sermons in 591 CE, Pope Gregory I conflated Mary Magdalene with the "sinful woman" in Luke 7:36–50 who anointed Jesus's feet with perfume and her tears, and dried them with her hair. Though the exact nature of the woman's sins were never revealed, many assumed she was either a repentant prostitute or at least enjoyed the company of men more than was socially acceptable. And suddenly she and Mary Magdalene were one and the same, and Jesus's beloved companion had been pornified and delegitimized for all time.

The stereotype of the sinful promiscuous woman was alive and well long before Mary Magdalene. Just think for a moment about Eve and that luscious apple, which evidently gave Adam the world's first human erection. Pandora, gifted with beauty and sensuality by the gods themselves. The adulterous Helen of Troy's breathtaking allure launching all those warships.

Delilah, using sexual wiles to seduce Samson into telling her about his hair. Jezebel, painting her face. Salome, dancing so erotically the salivating King Herod granted her wish to cut off John the Baptist's head and parade it around the feast on a party platter.

The purpose of slut-shaming is to silence women and usually has nothing to do with their actual sexual behavior. One powerful example is the story of Jezebel, whose name has become a byword for sluttiness. The only extant sources for her story were written by her most ardent enemies in the biblical books of 1 and 2 Kings. The daughter of the king of Tyre in the ninth century BCE, Jezebel was sent away from her balmy, sophisticated city on the Mediterranean coast probably at the age of fourteen or fifteen to marry the older King Ahab of Israel, who lived in the harsh desert of Samaria.

As was traditional at the time in the Middle East, the bride maintained her cultural identity by bringing her own deities with her to worship. Jezebel didn't seem to think much of the local god, an angry male deity named Yahweh who liked to smite people. She worshipped Baal and his consort Astarte, known as Asherah in Israel where the goddess had also been—perhaps among some people still was—the consort of Yahweh. While much is unknown, it seems that powerful forces among the Israelites were striving to cancel the historic female deity and transfer all power to a monotheistic male deity.

Entranced with his young wife, King Ahab allowed her to promote the worship of her gods, to create priests and build temples. Yahweh was furious and sent a three-year drought. Needless to say, it was all Jezebel's fault. Poor Ahab, who could have sent her home or locked her up in the harem, simply couldn't resist her wily ways. 1 Kings 21:25 says: "But there was none like unto Ahab, which did sell himself to work wickedness in the sight of the Lord, whom Jezebel his wife stirred up."

According to Kings, Jezebel arranged the judicial execution of a man named Naboth who refused to sell his vineyard to the king, falsely accusing him of cursing God. The elders and nobles of his town—with whom he was probably well acquainted—stoned him to death. But some biblical scholars believe the story doesn't make a great deal of sense. Janet Howe Gaines, a professor at the University of New Mexico specializing in the Bible, wrote, "If the trickster queen is able to enlist the support of so many people, none of whom betrays her, to kill a man whom they have probably known all their lives and whom they realize is innocent, then she has astonishing power. The fantastical tale of Naboth's death—in which something could go wrong at any moment but somehow does not—stretches the reader's credulity. . . . Perhaps the biblical compiler is using Jezebel as a scapegoat for his outrage at her influence over the king, meaning that she herself is being framed in the tale." (Another way to take down a powerful, ambitious woman is to accuse her of murder, as we shall explore in Chapter 11.)

Over time, resistance to Baal worship grew among the worshippers of Yahweh. The prophets of Baal and the prophets of Yahweh killed one another in massacres involving hundreds. They held magic contests to see whose god was more powerful. After the chief prophet of Yahweh, Elijah, won a major spontaneous-combustion contest and slaughtered 850 of the queen's prophets, she wrote him, "May the gods do the same to me and even more if tomorrow about this time I haven't made you like one of those prophets you had killed." Elijah was so terrified that he ran into the hills and hid. Clearly, this was a woman who had stepped out of the traditional bounds of an Israelite queen consort. She needed to be stopped.

Elijah was not the one who stopped her, though, as he was carried off the planet by something like a flying saucer, and no one ever saw him again. Jezebel must have been glad. His successor was Elisha, whose vanity caused him to murder dozens of children. When a group of small boys near the city of Bethel mocked him for being bald—evidently a sore point—he cursed them in the name of God and caused two bears to appear that tore forty-two of them to pieces. Then he decided to topple King Joram, Ahab's successor and the son of Ahab and Jezebel. Elisha crowned King Joram's top military commander, Jehu, the new king, and instructed him to kill his master.

When King Joram went out in his chariot to meet General Jehu, perhaps having heard some rumors of treachery, he asked him if he came in peace. Jehu retorted, "What peace, so long as the whoredoms of thy mother Jezebel and her witchcrafts are so many?" He then shot an arrow into the king's heart. In the context of the time and place, the words "whoredoms" and "witchcrafts" often meant idolatry, the pimping of oneself out to false gods, exotic statues, and incense-laden altars. In this sense, it had nothing to do with sex or witchcraft. And nowhere in a text that tries so hard to ruin Jezebel's reputation is she ever accused of being unfaithful to her husband. On the contrary, she is portrayed as being such a loyal supporter of Ahab's that she even murdered a respected citizen to get her husband the vineyard he wanted. Yet because of the word "whoredoms," Jezebel has become known as the Slut of Samaria.

The final nail in her slutty coffin came when she adorned herself to meet Jehu. Hearing that he was coming to slay her, she sat down at her dressing table, applied kohl to her eyes, arranged her hair, and bedecked herself as befitting a queen. Then she sat in her window to watch him drive up in his chariot. Readers throughout the centuries assumed that Jezebel, who was now a dignified widow and grandmother, was trying to seduce a much younger man who had just treacherously murdered her son—not very likely. Many biblical scholars, including Isaac Asimov in his two-volume *Guide to the Bible*, see Jezebel as insisting she meet her death on her own terms, imbued with royal grandeur, so Jehu understood exactly that he was killing a queen. Rather than showing her charms at the window, hoping to vanquish him through lust, she sat there calmly awaiting her murderer, looking down at him scornfully as he clattered up.

Calling to Jehu from her window, the mocking, insulting greeting she gave him clearly shows seduction was the farthest thing from her mind. "Have you come in peace, you Zimri, you murderer of your master?" she asked, comparing him to a former king who had also obtained his position by killing his rightful monarch. Jehu called up to her eunuchs, asking them if they were with him. In response, they flung her out of the window, and his chariot and horses trod on her, and her blood spattered the horses and the wall.

After eating and drinking in his new palace, Jehu thought better of leaving Jezebel's mangled body on the paving stones and issued orders for her to be buried, as she was a king's daughter. But his servants only found a few pieces of her: her skull, her feet, and the palms of her hands. Dogs had eaten the rest. "And the carcass of Jezebel shall be as dung upon the face of the field in the portion of Jezreel," the chapter ends, "so that they shall not say, This is Jezebel." In other words, the trollop had it coming.

Jehu, the righteous one and Yahweh's chosen, then ordered the murders of seventy boys related to Ahab—whose heads he piled in baskets on either side of the town gate—and all of Ahab's officials, supporters, their family and friends, and all the followers of Baal, creating great heaps of corpses of innocent men, women, and children, added to the forty-two children murdered by Elisha for calling him bald.

But the sins of Jezebel are what we remember.

CLEOPATRA'S IMPERTINENCE

Think hard for a moment about what you know of Cleopatra. That as a girl she was smuggled into the palace in a rolled-up carpet to meet Julius Caesar and sprang out on the floor ready to seduce the fifty-two-year-old battle-hardened general? That with her wanton ways and feminine wiles she persuaded Mark Antony to ditch Rome and loll around with her on silken sheets? That during the Battle of Actium with Rome, she selfishly sailed away to safety with her fleet, leaving Antony on his own? And when Octavian had her cornered, she killed herself with an asp?

These stories—that she was ambitious, selfish, and slutty—are straight out of the Misogynist's Handbook, written by her enemies, the Romans, to justify their conquest of a sovereign nation. (The only action of Cleopatra's the Romans did approve of was that she killed herself. As a reward for finally doing something right, her conqueror Octavian buried her in a grand mausoleum.)

A hundred and sixty years after her death, the poet Horace described her as "a crazy queen . . . plotting . . . to demolish the Capitol and topple the [Roman] Empire." Which was not at all true. The last thing on her mind was to bring Egyptian troops to Rome—which she must have considered lacking in every human comfort compared to the glittering sophistication of Alexandria—and conquer it. She just wanted Rome to leave Egypt alone. A century after Horace, the Roman poet Lucan branded her as "the shame of Egypt, the lascivious fury who was to become the bane of Rome." She was called a "harlot queen," "Ptolemy's impure daughter," "a matchless siren," the "painted whore" whose "unchastity cost Rome dear."

Some of the Cleopatra sex stories don't pass the laugh test. King Herod of Judea called her a "slave to her lusts," claiming after her death that she had tried to force herself upon him during a state visit to Jerusalem, an assault from which he had virtuously defended himself. Her enemies in Rome nicknamed her "Meriochane," which means in Greek "she who parts for a thousand men." She was accused of performing fellatio on a hundred Roman nobles in a day. The early third-century Roman writer Cassius Dio commented on Cleopatra's "insatiable sexuality." A late fourth-century writer, the Church Father Jerome, in a porno-tart fantasy a bit too heated for a man of the cloth, described her as "so insatiable that she often played the prostitute . . . so beautiful that many men paid for a single night with their lives."

In the thirteenth century, the Italian poet Dante consigned her to the second circle of Hell, where she joyously ruled as Queen of Lust. The fourteenth-century Italian writer Giovanni Boccaccio called her "the whore of eastern kings." Perhaps William Shakespeare, using earlier sources, played the greatest role in creating Cleopatra as we know her. In his classic play, he had the ambitious vixen wrapping poor, weak-minded Antony around her little finger. Undoubtedly, she had relationships with two powerful Roman generals—Julius Caesar and Mark Antony. But as far as we know, these were the only two sex partners she had in her life. Cleopatra's sin was not her sex life but the fact that she had more power and greater wealth than any Roman ever had up to that time. And that she was a woman.

Cleopatra became queen of Egypt upon the death of her father, Ptolemy XII, in 51 BCE, when she was eighteen years old. Though quite young to hold power, her father had provided her with an excellent education: rhetoric, science, history, economics, medicine, and fluency in nine languages, according to Plutarch, writing in the early second century CE. In addition to her native Greek, she spoke Hebrew, Arabic, Parthian, Ethiopic, and the local Egyptian language, which no other members of her Greek dynasty had ever bothered learning.

The queen was not considered classically beautiful. Plutarch noted that her beauty "was not in itself so remarkable that none could be compared with her, or that no one could see her without being struck by it." It was her charm, her wit, her intellect that made her "irresistible . . . bewitching."

She was a popular ruler, who effectively managed the economy and international relations. The Egyptian people considered her to be nothing less than a goddess incarnate. Her main concern was staving off annexation by the greedy, warlike Romans, the conquerors of many other lands who for some time had been eyeing the Nile breadbasket to feed their armies.

Her father had designated Cleopatra and her younger brother, Ptolemy XIII, joint heirs, and perhaps they even married—though in name only, as he was eight years her junior—as pharaonic brothers and sisters had done for centuries. But Cleopatra opted to rule alone, and Ptolemy, through his advisors, raised an army, forcing her into exile in Syria.

When Julius Caesar arrived in Alexandria in 48 BCE, Cleopatra decided to persuade him to keep her in power and somehow—in a sack of bedding, according to one Roman writer, which later morphed into the more romantic carpet—was smuggled into the palace to meet him. "It was by this device of Cleopatra's, it is said, that Caesar was first captivated, for she showed herself to be a bold coquette," according to Plutarch, writing in about 100 CE. "And it was there that the young Ptolemy XIII found them early the next morning, aghast that Caesar already had been seduced by his half-sister."

The fact is that no one knows if the twenty-one-year-old Cleopatra had sex with the fifty-two-year-old Julius Caesar that very night—according to Roman sources she did, the little tramp, but how could they actually know? And it is laughable to think that the battle-hardened general who had conquered Gaul became putty in her soft, devious hands, completely enslaved at a glance by this not terribly pretty woman. However, Cleopatra and Caesar certainly did begin a romantic relationship at some point. When Ptolemy XIII battled Cleopatra for sole control of the throne, Caesar sprang to her defense, and Ptolemy drowned while trying to escape. Caesar put Cleopatra firmly back in power. When the poet Lucan wrote, "Cleopatra has been able to capture the old man with magic," the magic was sex, yes, but not sex alone. Caesar could have gotten tantric sex anywhere with just about anyone. He must have been entranced by the only woman he had ever met who was his equal. At twenty-one, she had raised armies, controlled a complex economy, dispensed justice, and made treaties with foreign powers.

We don't know how Cleopatra felt about Caesar. Was it love? Admiration? Or a strong instinct for survival? Without Caesar's assistance, it is fairly certain that Cleopatra would have been dead or exiled in her early twenties. Though Rome would accuse the queen of overweening ambition, perhaps her greatest ambition was to live.

Cleopatra gave birth to Caesar's son, Caesarion, after he returned to Rome, and in 46 BCE visited him there, holding court at one of his villas. Roman noblemen, whose ideal woman was a chaste and humble mute who spent her time spinning cloth, were aghast at a powerful, crazy-rich, independent foreign woman—the acknowledged mistress of their top general—flamboyantly sweeping through the forum with a trail of servants swinging incense burners. The Roman statesman and orator Marcus Tullius Cicero called the queen of Egypt "impertinent." (How can a queen be impertinent?) Her behavior "made my blood boil to recall. . . . I hate the queen!" he wrote. He sniffed, "Her way of walking . . . her clothes, her free way of talking, her embraces and kisses, her beach-parties and dinner-parties, all show her to be a tart."

The tart returned to Egypt after Caesar's assassination, and the Roman Republic dissolved into a brutal civil war as Caesar's supporters battled his killers. The supporters won, and three years after the Ides of March, Caesar's heir and great-nephew, Octavian, controlled the western Mediterranean while his colleague Mark Antony controlled the east. Antony summoned Cleopatra to meet him in Tarsus, on the coast of what is now Turkey. She sailed up in a gilded barge, with billowing purple sails, clouds of incense wafting before her as she reclined on a couch, her servants dressed as nymphs and cupids. And so began her second romantic relationship to save her country from Roman annexation. She would have three children with Antony.

The burly, bearded Roman general was smitten with both Cleopatra and her opulent lifestyle. He neglected his long-suffering wife, Octavia, who was Octavian's sister, and spent undue amounts of time in Egypt rather than going about Rome's business. His goal was for greater power and wealth than he could ever possess in the comparative backwater of Rome: he was positioning himself as emperor of the Eastern Empire. In 34 BCE, rather than annexing large swathes of newly conquered territory as Roman provinces—which all good Roman generals proudly did—Antony declared Cleopatra's children monarchs over Armenia, Media, Parthia, Cyrenaica, Libya, Syria, Phoenicia, and Cilicia.

When Antony divorced Octavia in 32 BCE, it was the last straw for her brother. Hesitant to declare war on Antony, still a popular general, Octavian had the brilliant idea of declaring war on Cleopatra, the degenerate foreign woman who through sorcery had unmanned even the most virtuous Roman, Mark Antony, causing him to forget his duty, drop his sword, and dally in her fine, smooth arms in a perfumed haze of oriental inertia. Even when he managed to clamber up on a horse, he was worse than useless. The fact that Antony had lost the war with the Parthians in 36 BCE, along with twenty-four thousand men, was all *her* fault.

During the naval battle of Actium five years later, Cleopatra sailed off with her fleet of sixty ships, followed by Antony, as the fighting continued without them. It could have been a plan prearranged with Antony or confusion in the midst of battle, yet Rome portrayed it as the queen saving herself and leaving him to his fate. She was *untrustworthy*.

When Octavian invaded Alexandria, Antony, cornered, killed himself. Cleopatra, vowing never to march in a Roman triumph, did the same, reportedly dying from the bite of an asp she had smuggled into the tomb where she had been holed up. But no one really knows. According to Plutarch, Cleopatra was known to wear a hollow comb in her hair filled with poison. Perhaps she simply drank poison concealed in her hair comb, though that story isn't nearly as dramatic as holding a viper to one's breast. It is likely that the snake, a prehistoric symbol of female power (discussed in-depth in Chapter 14), represented Cleopatra's agency in avoiding the shame of being paraded in a Roman triumph.

After Octavian defeated Cleopatra and conquered Egypt, "Validity was restored to the laws, authority to the courts, and dignity to the senate," proclaimed the historian Velleius Paterculus, writing a century after her death. The powerful woman, who upset the natural order of things, was gone, the power rightfully returned to men. The world was safe once more. Egypt would not become a truly sovereign nation again until 1953.

A notable twentieth-century British historian of Cleopatra's, W. W. Tarn, wrote that against her "was launched one of the most terrible outbursts of hatred in history; no accusation was too vile to be hurled at her, and the charges which were made have echoed through the world ever since and have sometimes been taken for facts." To destroy Cleopatra's reputation for all time, and to keep other ambitious women in their place, the Romans reduced the queen's successful statecraft to sex, insisting her power existed not between her ears, but between her legs.

MESSALINA'S GILDED NIPPLES

For nearly two thousand years, the name Messalina has been synonymous with rapacious sexual promiscuity. Given that the Roman propaganda machine documenting the empress's depraved sex life was simultaneously beating up Cleopatra for similar crimes, it is hard to say how much is true about Messalina. Perhaps the woman was indeed a murderous nymphomaniac; there must have been one or two such women sprinkled throughout history, after all. But the stories about her seem more like a misogynistic sex fantasy.

Born into the scorpion's nest of the Roman imperial family between 17 and 20 CE, at the age of about eighteen Messalina married her first cousin once removed, the future emperor Claudius, who was forty-seven. They had two children, a girl and a boy. According to later Roman sources, Messalina, once empress, persuaded Claudius to execute or exile female relatives who aroused her jealousy and tried to have the young Nero murdered in his bed as a rival to her own son for the imperial throne.

She killed her stepfather, Appius Silanus, because she wanted him for herself and couldn't have him. And, like Jezebel before her, Messalina had another man, Valerius Asiaticus, murdered because she coveted a plot of his land. She terrorized Poppaea Sabina the Elder, a rival, until the poor thing committed suicide. She poisoned Marcus Vinicius because he refused to sleep with her. But it wasn't her supposed murders—which were fairly common in her family, after all—that destroyed her reputation for all time. It was her sex life.

Some seventy years after Messalina's death, the Roman historian Suetonius wrote, "To cruelty in the prosecution of her purposes, she added the most abandoned incontinence. Not confining her licentiousness within the limits of the palace, where she committed the most shameful excesses, she prostituted her person in the common stews, and even in the public streets of the capital."

Writing around the same time, the satirist Juvenal called Messalina "the imperial whore," just as the poet Propertius had called Cleopatra "the harlot queen." Juvenal wrote, "Hear what Claudius had to endure. As soon as his wife perceived he was asleep, this imperial harlot, that dared prefer a coarse mattress to the royal bed, took her hood she wore by nights, quitted the palace with but a single attendant, but with a yellow tire [wig] concealing her black hair; entered the brothel warm with the old patchwork quilt, and the cell vacant and appropriated to herself. Then took her stand with naked breasts and gilded nipples, assuming the name of Lycisca, and displayed the person of the mother of the princely Britannicus, received all comers with caresses and asked her compliment, and submitted to often-repeated embraces."

Pliny the Elder wrote that Messalina staged a competition with another prostitute, Scylla, to see who could have sex with the most men in a twenty-four-hour period. According to Pliny, Messalina won with twenty-five men, long after the real whore had given up in exhaustion. Did this actually happen? Or was it a game of historian one-upmanship to see who could most thoroughly destroy the woman's reputation? If so, Pliny won, creating a male sex fantasy for the ages.

The strangest, most head-scratching story of all about Messalina was that when Claudius was out of town, she bigamously married her lover, Senator Gaius Silius, in Rome and held a lavish wedding banquet. Why, as empress, with as many lovers as she wanted and a dim-witted husband who didn't suspect a thing, would Messalina make such a dangerous marriage? Roman historian Tacitus wrote it was because she had "become sated with the simplicity of her adultery" and wanted something else to satisfy her insatiable lust. Which makes no sense whatsoever. At any rate, hearing of her treachery, Claudius raced back to town. Not knowing what to do with his unfaithful wife, he decided to put off any punishment until the following morning, but his servant, who despised her, stabbed her in the night. And when Claudius heard about her execution the next day, he merely called for wine. Then the Roman Senate ordered a *damnatio memoriae* so that Messalina's name would be removed from all public and private places and all statues of her would be taken down. The entire disjointed story does not add up.

The earliest tales of Messalina that have come down to us were written at least seventy years after her death by three Roman writers: Tacitus, who admitted his account seemed exaggerated; the gossipy Suetonius; and Juvenal, who wrote satire, and all of whom lived in a political environment hostile to the imperial line Messalina had belonged to. Clearly, something bad had happened with Messalina. Perhaps she had tried to take power, or maybe she already had it due to her influence over Claudius, which infuriated men in the imperial circle who vowed to get rid of her and control the emperor themselves. Perhaps her policies conflicted with those of Claudius's chief ministers and in a fit of temper she threatened one of them, as happened with Anne Boleyn and Thomas Cromwell. Whatever occurred, it is ridiculous to believe Messalina snuck out of the palace and stayed away all night without her husband noticing, even when she carried "home to her Imperial couch the stink of the whorehouse," according to Juvenal. Or that she thought she could get away with a very public bigamous marriage and her husband wouldn't notice that either.

In a 2011 article in *Constructing the Past*, historian Kristen Hosack concludes that "the descriptions of Messalina in the works of Tacitus, Suetonius, and Juvenal are exaggerated, invented, or intentionally misleading, which means that they are not entirely accurate representations." Or perhaps, as with many other women represented in this chapter, they are not accurate at all.

WHEN EMPRESS THEODORA'S PRIVATE PARTS APPEARED ON HER FACE

Five centuries after Messalina gloriously gilded her nipples, the Byzantine empress Theodora—beautiful, intelligent, and powerful—also inspired male fantasies of gross sexual impropriety. In her case, she had a truly shady past, raised as she was in a family of entertainers at the raucous Constantinople Circus, where chariot races and gladiator fights were held. She became a well-paid courtesan, but in her early twenties she gave up the profession for something more respectable, if far less remunerative: working wool. The emperor's heir, Justinian, fell in love with her and married her, despite her background, and at the age of twenty-five, she became empress, wielding great influence over her husband's political and religious affairs. The contemporary historian Procopius wrote that she was a raging nymphomaniac, an accusation that does not accord with her choice to spin wool instead of acting as mistress to rich men.

"Often she would go to a bring-your-own-food dinner party with ten young men or more," he wrote, "all at the peak of their physical powers and with fornication as their chief object in life and would fornicate with all her fellow-diners in turn the whole night long. When she had reduced them all to a state of exhaustion, she would go to their menials, as many as thirty on occasion and copulate with every one of them, but not even so could she satisfy her lust." Here was a woman, according to Procopius, who could out-copulate Messalina.

He continued (and we can picture him rubbing his hands or perhaps something else with glee), "Though she brought three openings into service, she often found fault with Nature, grumbling because Nature had not made the openings in her nipples wider than is normal so that she could devise another variety of intercourse in the region."

The historian describes one of her theater acts where she lay almost naked on the ground, servants sprinkled grain on her private parts, and geese would nip them off. As if that weren't enough, he insisted that due to her years of sinful lust God made her private parts appear on her face, and that, according to several reliable witnesses, Theodora was a demon whose head would leave her body and go roaming around the palace at night.

Procopius's spite probably stemmed from the fact that

Theodora—a woman, and a low-born former prostitute, at that— wielded unlimited power granted by her husband. Indeed, she was arguably the most powerful woman in the history of the Roman empire. Nor did she forget less fortunate women, as she herself had once been. She bought hundreds of young women who had been sold into prostitution, freed them, and made sure they had legitimate means of earning a decent living. She closed brothels and arrested pimps. The generous empress was as different from a disembodied demonic head with two vaginas for eyes and a clitoris for a nose as she could possibly have been.

"THE MOST IMMORAL WOMAN OF HER AGE"

Isabeau of Bavaria is perhaps best known today as the mother of Charles VII, the weak French king whom Joan of Arc encouraged to fight invading English forces. But Isabeau is also known for many other things: treason, profligacy, political incompetence, greed, and, most of all, adultery with her brother-in-law, Louis d'Orléans. If we scrape below the sexist varnish of her story, we find none of this to be true. Here is yet another queen carved into a grotesque shape by misogyny alone.

A Bavarian princess, in 1385 fourteen-year-old Isabeau married seventeen-year-old King Charles VI of France, who fell in love with her at first sight. They lived happily for seven years until one day, riding through the woods, the king suffered a murderous fit of what was probably schizophrenia—and killed four members of his entourage before others chained him up. When he recovered several days later, he had no memory of what had passed. The fits came and went without warning, some lasting for months. During his bouts of madness, Charles was unaware that he was king. He didn't recognize his wife. When Isabeau attempted to soothe him, he struck her, hurled both obscenities and objects at her, and asked his attendants, "Who is this woman obstructing my view? Find out what she wants, and stop her from annoying and bothering me, if you can." At her wit's end, the queen finally appointed a lovely young mistress to keep him calm and occupied.

During his periods of temporary insanity—which came to be known as his "absences"—sometimes Charles believed he was made of glass and could easily shatter; he instructed his tailor to put iron rods in his clothing to prevent him from breaking. He sat stone still for hours, afraid of cracking and falling onto the floor in a heap of glass shards. Other times, clearly not worried about shattering, he raced through the palace howling like a wolf and cavorted nude in the palace gardens.

With the king so often unable to rule, his uncle, Philippe of Burgundy, and his brother, Louis d'Orléans, fought for control of the kingdom and its treasury. It was a late-medieval French version of the Hatfields and the McCoys, both feuding sides with armies behind them burning, raping, and murdering. When Philippe of Burgundy died in 1404, the nation must have hoped the deadly rivalry was at an end. Then his son, Jean sans Peur (Jean the Fearless), continued to fight his cousin Louis d'Orléans with even greater vigor, culminating in Jean's murder of Louis in 1407.

Each time the king woke from his madness, he asked the queen what had happened, acted on her advice, and took back the reins of power. In 1402, evidently fearful one of the royal rivals would take

total control of him, he appointed Isabeau as official royal media-
tor between them. Year after year, with her insane husband howling like a wolf, Isabeau did what she could to promote peace and
save the kingdom for her son. She negotiated, persuaded, bribed,
charmed, and intervened, doing an impossible balancing act between forces that did not want to compromise, that would gladly
have consented to the total destruction of the country if it served
their personal interests.

The bond between Isabeau and her husband was close when
he was sane. She often went on pilgrimages to shrines where she
prayed for her husband's recovery. She had seven of her twelve children after his attacks started, and over the years he entrusted her
with more and more power, appointing her leader of the regency
council; guardian of the heir to the throne, the dauphin; and giving
her control of the royal treasury. In 1408, Charles announced that
Isabeau would preside over the government of Paris in his "absences." Despite her unceasing efforts for peace, full-blown civil
war broke out in 1411.

In 1415, the dauphin, Louis, turned eighteen. Determined to
find a peaceful solution to the feud, and just as he was playing a
larger role in French politics, Louis died of dysentery. Fourteen
months later the next-eldest son, Jean, died at nineteen of an abscess on the head. Isabeau's youngest son, the unappealing Charles,
became dauphin. In 1419, the sixteen-year-old signed a peace
agreement with Jean sans Peur and agreed to meet on a bridge two
months later for another discussion. On the bridge, Charles treacherously had Jean hacked to death with an axe right in front of him,
revenge for his killing of Louis d'Orléans back in 1407.

When Charles VI learned of the murder, he disinherited his son for once again setting in motion the feud that had been ravaging the nation for twenty years. His new heir would be King Henry V of England, who had taken advantage of France's internal chaos and begun an invasion four years earlier. Henry now had total control of Normandy, and his forces threatened the rest of the country. He demanded the hand of Charles and Isabeau's daughter Catherine and the crown for himself after Charles's death. Charles and Isabeau believed that the Treaty of Troyes, as their agreement was known, would end not only the war with England, but also the destructive, decades-long feuding among French noble families. And if Henry and Catherine had a son—the half-French grandson of Charles and Isabeau—he would be king of France.

Historians of later centuries saw the handing over of the French crown to an English invader as a shocking betrayal of France. At the time, however, much of France accepted the treaty with hope; it could solve the unsolvable. As it turned out, both Charles VI and Henry V died in 1422, Henry leaving an infant half-French son with Catherine. But most Frenchmen feared a long regency—with yet more uncles and cousins fighting for power—and considered the Treaty of Troyes null and void. It had, after all, been signed by a crazy man. Better to get behind Charles VII as king.

It is, at this point, probably not surprising that with all the evidence of men behaving badly—burning crops and villages, raping women, stealing from the national treasury, assassinating one other, and running around naked howling—it is the woman who comes out of the story as the villainess, the woman who, married to a lunatic, devoted her life to pacifying warring factions to save the kingdom and protect her children. Yes, she had been friendly with her husband's brother, Louis d'Orléans, as they both tried to find cures for the king's madness. She sometimes sided with him against the more frightening Burgundians, though she worked with both sides to end the feud. But what had she done to become known forever as the harlot queen of France?

In 1405, a chronicler named Michel Pintoin, a supporter of the Burgundians who were, at that moment, angry with the queen for limiting their access to the royal treasury, wrote about a sermon given to the court by a monk named Jacques Legrand. "Lady Venus occupies the throne in your court," Pintoin quoted Legrand as saying, "certainly drunkenness and debauchery follow her, turning night into day, with continual dissolute dancing." Pintoin then reported on Legrand's criticism of "the dissoluteness of their clothing, of which the queen had been a principal instigator and whom he reproached in many ways." Finally, Pintoin had the monk bellowing, "This and many other things, oh Queen, are said about your court to its disgrace."

Preachers have been fomenting about dancing, immoral behavior, and immodest clothing since the dawn of Christianity, especially at royal courts. It would be impossible from this text alone for anyone to believe that the queen was having an affair with her brother-in-law. Pintoin's report of general bad behavior at court was most likely propaganda planted by the duke of Burgundy in an effort to blacken his enemies' reputations and gain more power. The second contemporary document, an anonymous pamphlet called the *Songe Veritable* (the Genuine Dream), was also a Burgundian propaganda piece, attacking Louis d'Orléans and his supporters. Queen Isabeau is criticized merely for "getting everything she can." No adultery is mentioned, which it certainly would have been if rumors had been around at the time.

How, then, did the black legend of Isabeau of Bavaria start? Jean Chartier, appointed royal chronicler by Charles VII in 1437, wrote that a few years earlier the English had spread the tale that Isabeau's son Charles was not the son of Charles VI. The rumor was around by 1429, when the widowed Isabeau was living quietly in retirement. It seems the English, feeling threatened by Joan of Arc's victories, told tales of Isabeau's infidelity and Charles VII's bastardy to strengthen their claims on the French throne.

Then, in 1791, the French revolutionary writer Louise-Félicité de Kéralio wrote a book in which she presented Isabeau as a prototype of Marie Antoinette. Isabeau, draped in diamond necklaces, cavorted with her brother-in-law (as Marie was accused of doing), cared nothing about the suffering of the people, and said "let them eat cake." Then the slut sold the country out to the English, just as Marie was said to have betrayed France to the Austrians.

Historians of the nineteenth century took the villainess story and ran with it. They pointed accusingly to language in the Treaty of Troyes—which Isabeau agreed to and her husband signed— referring to the dauphin Charles's murder of Jean sans Peur. Charles was being disinherited, the document made clear, for "the horrible and enormous crimes perpetrated upon the kingdom of Grace by Charles, the so-called dauphin." At the time, the use of the word "so-called" (in French, *soi-disant*) was a common insult, a sneer at the person as unfit for his office. Yet later historians saw it as proof that Isabeau was making known her son's illegitimacy. It is ridiculous to imagine that the king branded his wife a trollop—and she happily went along with it—to get their problem son off the throne.

By the twentieth century, Isabeau's adultery and profligacy were accepted as facts. In 1903, French historian Marcel Thibault, extrapolating heavily from the monk's sermon, wrote in his *Isabeau de Bavière: Reine de France*, "She did not try to stop Charles VI, engaged in a downward spiral of pleasures. . . . She lived in a whirlwind of insane amusements and splendid celebrations. And while the king wasted his strength, compromised his dignity, ruined his intelligence, she, because of her immoderate lifestyle, produced for the kingdom only sickly babies."

She neglected her children from the moment of conception, not caring if they were born sick (though chroniclers of the time and Charles himself in his official edicts noted her devotion to her children; she often raced with them out of the way of threatening armies). She hated her husband (though Charles assigned her to supervise his treatment during his "absences," not trusting anyone else). Why limit her to one lover? She had countless! She was politically incompetent, jumping from one side to the other during the feud (to keep a balance of power to protect the realm).

As if that weren't enough, somehow, in the centuries after her death, Isabeau became fat, so fat that palace doors had to be widened for her to pass through. "At the end of her life," according to Mary Gordon's 2000 book *Joan of Arc*, "she was grotesquely fat, to the point that her obesity made it doubtful that she could act as regent of the kingdom." (She was too fat to rule, though being such a frivolous person, she could still dance.) Another historian, Jean Verdon, in his 1981 book *Isabeau de Bavière*, blamed her political failures after the deaths of her two older sons on her obesity. "Only Isabeau remained. But overweight, and perhaps distressed by the deaths of her two sons, she seems not to have played an important role in political life."

As an ugly, fat adulteress who despised her own children, who feasted, danced, and wore expensive clothing while the people starved, and a traitor to the nation to boot, clearly this woman was a villainess beyond redemption. Isabeau was "the most immoral woman of her age," according to a 1992 book, *The King's Women*, by Dinah Lampitt.

When historian Tracy Adams began researching Isabeau for her 2010 book, *The Life and Afterlife of Isabeau of Bavaria*, she initially believed, "Surely where there is smoke there is fire," as she wrote in her introduction. Adams dug around in contemporary documents and came up mystified. "When I tried to verify the charges," she wrote, "I could not. The histories . . . do not cite contemporary references, but each other, in a seemingly endless feedback loop. Moreover, their favorite adjectives for describing the queen— wanton, cupidinous, obese—give pause, representing as they do a litany of traditional misogynistic complaints. Claims that on the surface suggest misogyny deserve skepticism."

Adams found, "The difference between the Isabeau I discovered there and the promiscuous creature of the more general histories was nothing less than astounding. To begin, the few scholars who have studied her in any details are unanimous in their assessment that the adultery charge is a fantasy, resulting primarily from the misreading of two documents dating from the queen's lifetime."

Adams found more recent research that has tried to rehabilitate the queen's reputation. In the 1970s, French historian Yann Grandeau praised the queen's "acute intelligence." "Directing the game from the sidelines and giving an impression of serenity to all," he wrote, "she followed a centrist political line with foresight and tenacity." Historian Rachel Gibbons credited her with preserving the monarchy, praising her "often adept handling of diplomacy."

As the award-winning historian Ronald Schechter wrote in 1998, "Once a fact becomes 'common knowledge' the historian is released from the obligation to cite a source, and only a determined effort to falsify it can dislodge it from the bricolage of generally accepted facts that constitute the historical canon."

And so, Isabeau of Bavaria has gone down in history like this:

She was a slut.

She was a bad wife.

She was a bad mother.

She was ambitious.

She was fat.

She was frivolous.

She spent too much money on clothes.

She was greedy.

She was unlikable.

She was untrustworthy.

And probably, her voice was shrill.

"A STABLE OF WHORES"

In the sixteenth century, the powerful queen mother of France, Catherine de Medici, was *too* powerful by many estimates. She needed to be taken down. But some women just didn't fit the mold of sexy seductress or brazen whore, no matter how much their enemies wanted to shoehorn them into it. By the time she wielded power, Catherine was forty-two and obese from giving birth to ten children. Even in her youth, she had never been beautiful or seductive, though ambassadors described her as intelligent and charming. So when her male detractors were itching to throw the *W* word at a stout middle-aged widow swathed in black and clanking with rosaries, they just couldn't. No one, hearing such an accusation, could have kept a straight face.

Her enemies came up with the next best thing: she was a madam, sending her gorgeous ladies-in-waiting to seduce powerful men, squeeze information out of them, and report back to her. Catherine wasn't a whore herself—of course not, just look at her—but she ran "a stable of whores," also called her "flying squadron" because they leapt provocatively into the air when dancing. Research, however, indicates she ran an orderly household of mostly respectable young women of good families.

The contemporary historian Pierre de Bourdeille, abbé de Brantôme, who spent years at the court and was well acquainted with all the characters, wrote in his *Book of the Illustrious Dames* of the love affairs of a few of Catherine's ladies-in-waiting, a thing hardly shocking at a royal court. (The stern Elizabeth I threw some of her ladies into the Tower for such crimes against maidenly virtue.) But Brantôme also wrote of Catherine, "She had, ordinarily, very beautiful and virtuous maids of honor, who conversed with us daily in her antechamber, discoursing and chatting so wisely and modestly that none of us would have dared to do otherwise; for the gentlemen who failed in this were banished and threatened." Catherine de Medici's stable of whores was a misogynistic fantasy created by a cadre of angry men to punish the queen mother and her ladies-in-waiting for acquiring increasing prominence on the national stage.

Catherine was accused of pimping in the contemporary book *A Mervaylous Discourse*. The anonymous author wrote that to gain control over Antoine, king of Navarre, "She entertained him to her power in all courtlike pleasures, in so much that he craving the favor of one of her ladies named the lady de la Rouhet, herself commanded this said lady not to refuse him of any requests which he might make unto her." That the lady de la Rouhet bore the king's son the following year was proof to many that the evil queen had pimped out her lady-in-waiting.

The *Discourse* accuses Catherine of draining all ambition from her son, King Charles IX, by encouraging him to live a life of reckless pleasure. She raised him to "haunt nothing but cockpits, even so now she endeavoreth to corrupt his youth causing him to be beset by bawds, whom she placeth next unto his own person." Her goal was to keep all the power to herself by "causing of him to forget his own affairs and to the making of him drunken in all delights . . . for she so lulled him in these aforesaid pleasures, that he never came unto counsel but by importunate suit of diverse gentlemen who misliked his lewd bringing up." And with her son loafing about in bed with lovers, all the power would be held in Catherine's manipulative little Italian fingers.

Across the English Channel, Elizabeth I was loved and revered as the Virgin Queen in her own realm, but her power and confidence outraged fellow monarchs, especially the tradition-bound King Philip II of Spain. Rumors—many of them spread by Spain—swept across Europe as to why she never married. She was frigid. She was a nymphomaniac and could not limit her lusts to one man. She only liked Black men, of which there were few in England, and she could not have married one of them anyway. The prize for the most creative theory goes to the Venetian ambassador to France, who wrote that the queen suffered a birth defect in her reproductive organs, which caused her to menstruate out of her left foot into her shoe (which must have been quite a mess).

The most persistent rumors of the queen's promiscuity concerned the favorite of her early years on the throne, Robert Dudley, earl of Leicester. It is possible their love was fully consummated, though in 1562, when Elizabeth thought she was dying from smallpox, she swore that it had not been. Back then, when a moribund individual fully expected to meet God at any moment and be sentenced to hell for lying, a deathbed oath was a kind of lie-detector test. Foreign courts, however, hearing the queen had been recuperating from an "illness" for several weeks, assumed it was an excuse to give birth secretly to Dudley's child. Even though the child would be illegitimate, European monarchs were eager to arrange a marriage for a child of theirs with the person who might be heir to Elizabeth's throne. In 1575, the bishop of Padua heard "that the Queen had a daughter, thirteen years of age, and that she would bestow her in marriage to someone acceptable to His Catholic Majesty [Philip II of Spain]." When Elizabeth's counselors were asked about betrothing this nonexistent daughter, however, they howled with laughter. It is likely that the queen didn't think it was very funny.

The Spanish ambassadors to England generally disliked the Protestant queen who had rejected the marriage proposal of their most august master. One of them, Don Diego Guzmán de Silva, upon arriving at his post in 1564, endeavored to find if there was any truth to the stories about her sexual promiscuity. Try as he might, he could not. Elizabeth herself told him, "I do not live in a corner. A thousand eyes see all I do, and calumny will not fasten on me forever." On another occasion, hearing the sexual slander about her being bruited about abroad, she said, "My life is in the open, and I have so many witnesses. I cannot understand how so bad a judgement can have been formed of me."

And, in truth, the queen was always attended by seven ladies of the bedchamber, six maids of honor, and four female chamberers. Several ladies slept in her bedroom each night. They bathed her, dressed her, and assisted her on the chamber pot. It is safe to say she was rarely, if ever, alone. It was clear to all, though, that the queen reveled in flirtations with handsome courtiers, with whom she surrounded herself. Perhaps she felt no need to hide her appreciation of male beauty simply because there was nothing to hide.

In 1565, Austrian imperial ambassador Adam Zwetkovich made "diligent enquiries concerning the maiden honor and integrity of the queen." He, too, was unable to find any evidence at all and decided the stories were but "the spawn of envy and malice and hatred."

Her subjects, certainly, believed in her chastity. In 1573, when a man in the Yorkshire town of Doncaster shouted slanderous sexual accusations about the queen in the street, a mob would have torn him limb from limb if the local magistrate had not intervened.

In 1572, Catherine de Medici, who well knew the viciousness of sexual slander, discussed the subject of Elizabeth's reputation with the English envoy: "And I told him it is all the hurt that evil men can do to noble women and [female] princes, to spread abroad lies and dishonorable tales of them, and that we of all princes that be women are subject to be slandered wrongfully of them that be our adversaries, other hurt they cannot do to us."

The most ingenious story about Elizabeth's sexuality is that she was actually a man, an imposter quickly swapped for the real Elizabeth after she died young. The theory was put forth by author Bram Stoker in the nineteenth century based on old village lore. This, at least, would satisfactorily explain to misogynists why Elizabeth's reign was the greatest ever.

CATHERINE AND THE HORSE

Catherine the Great so transgressed the boundaries set by the Patriarchy that it retaliated by creating the infamous lie that she had died at the age of sixty-seven while trying to have sex with a horse, which, being cantilevered into a promising position, had fallen and crushed her. What had she done to deserve such a takedown? As an impoverished German princess, married to the sadistic, pockmarked heir to the Russian throne at the age of sixteen, she was supposed to have children, stay in the background, and shut the hell up. Instead, Catherine rebelled against her crazy husband, who was planning to kill her, and seized the crown for herself.

Her success as a ruler horrified misogynists. Upon taking the throne, she found an empty treasury, 200,000 peasants on strike, a restless army unpaid for eight months, rebellion across the empire, unfathomable corruption in the legal system, and a near paralysis of commerce. Working fourteen hours a day, Catherine undertook reorganization of almost every aspect of Russian government with traditional German efficiency, presiding personally over council and senate meetings, peppering officials with probing questions they could not answer, and prolonging their working hours.

Within a year, she had founded an orphanage, a school for midwives, an organization for public health, and a school for the daughters of the nobility. She invited to Russia doctors, dentists, engineers, craftsmen, architects, gardeners, artists, and, of course, her favorite philosophers. She threatened male European monarchs both with her armies and her stream of invitations to the intellectual and artistic elite to leave their home countries and bestow their gifts on hers. England, France, and Prussia feared her.

To destroy her reputation and put her in her place, her enemies focused on her serial monogamy. Every few years, Catherine would choose a new lover and remain faithful to him until they broke up. Sometimes the lover ended the relationship in order to marry; sometimes she ended it if he bored her intellectually. Whatever the case, she refused to play the hypocrite and lie about her sex life.

In her memoirs she wrote candidly, "Nothing in my opinion is more difficult than to resist what gives us pleasure. All arguments to the contrary are prudery." Catherine thought society's preoccupation with female chastity ridiculous and laughed at the stories of her nymphomania and palace orgies. Though she enjoyed sex, her work took precedence. "Time belongs not to me, but to the Empire."

She never permitted off-color jokes in her presence and maintained a strict decorum at her court. She gave in to her passions at night behind tightly closed doors. But here was something for her enemies to seize upon. They didn't even have to make up lies out of whole cloth about her sex life as they did for other powerful women; they just needed to wildly exaggerate the truth. For instance, when her lover Alexander Lanskoy died in 1784 from diphtheria at the age of twenty-six, people said she had exhausted him with her sexual demands; he had died in her bed, while trying in vain to satisfy her insatiable passion. She had forced him to swallow poisonous aphrodisiacs, potions so strong they had made his body swell up and burst. The arms and legs fell off of his corpse. Newspapers across Europe published crude cartoons and sexually explicit tales about her.

She committed one other crime against the Patriarchy. Like Egyptian queen Hatshepsut 3,300 years before her, the empress refused to step aside when the male heir—her mentally unbalanced son Paul—came of age. The horse story was a posthumous punishment for her power, her self-confidence, her independence, and, most of all, her brilliant success, savaging her reputation as a wise and benevolent ruler for all time.

THE UTERINE FURORS OF MARIE ANTOINETTE

In 1770, when fourteen-year-old Austrian archduchess Marie Antoinette married the fifteen-year-old dauphin, Louis of France, her one job was to produce royal babies. But the lovely blond bride terrified Louis, awkward and overweight, who found himself unable to consummate the marriage. As the years passed, word of the stalemate got out and was picked up by the underground press. For more than a century, muckraking pamphleteers—called *libellistes*—had published vicious pamphlets—called *libelles* or *chroniques scandaleuses*—attacking the rich, royal, and famous, painting aristocratic debauchery in loving detail while simultaneously condemning it.

It was treason to libel a king—and even the most hardened pamphleteer seemed hesitant to do so—but anyone else was fair game. The queens of previous reigns, however, had been quiet, plain, pious women, content to bear children, do needlework, and pray. These meek souls, obediently accepting the dictates of the Patriarchy, were not deserving of the handbook's punishment. Therefore, the pamphlets had a field day excoriating the king's mistresses: pushy, greedy women who, by virtue of their very position, admitted to sinful sexual behavior.

But Louis XVI—who became king in 1774—didn't have a mistress. And his wife was a gorgeous foreign woman he could not satisfy in bed. The possibilities for lewd stories and the profits they would bring were endless. Additionally, the queen didn't stay silently in the background, praying and embroidering, as had her predecessors. The glamorous, much-talked-about leader of fashion, she was more like a royal mistress than a queen. She had the beauty and visibility of Louis XIV's demanding, tawny-haired lover, Madame de Montespan, and the exquisite taste of Louis XV's Madame de Pompadour. But those women, whatever their sexual behavior, were, at least, French. Like Catherine de Medici two centuries earlier, and Isabeau of Bavaria four centuries earlier, Marie was a foreign queen of France who stepped outside traditional boundaries. As a result, she soon found herself at the dangerous crossroads of xenophobia and misogyny.

In the fall of 1775, the first pamphlets circulated making fun of the king's impotence and regaling readers with the numerous ways the queen found sexual satisfaction. Her close friend and lady-in-waiting, the princesse de Lamballe, used her "little fingers," according to one. Several accused the queen of having an affair with Louis's younger brother, the dashing Charles-Philippe, comte d'Artois, with whom she was known to be close.

In 1776, Marie wanted to watch the sun rise over the Versailles gardens with a party of friends. Louis, who loved to sleep late, permitted it, as long as she didn't expect him to attend. Accompanied by the dour mistress of the queen's household, the Comtesse de Noailles, and several bodyguards, the queen and her friends enjoyed watching the sky slowly light up, seeing the first rays of the sun catch on the golden stone of the palace. Soon after, a pamphlet addressed to her personally described how the queen had used the occasion to run into the bushes for sexual encounters. In the public's mind, her innocent adventure had become an orgy.

Marie didn't care about the pamphlets. As a proud Habsburg princess and queen of France, she believed herself to be impervious to such drivel and considered it beneath her attention. But the king was furious and as impotent to stop their spread as he was in bed. In France, printers needed an official license to publish. But many printers lived in London or Amsterdam and smuggled the *libelles* into Paris. Footmen, for instance, were often bribed to hide the bundles in the trunks of noblemen returning to Paris who were too aristocratic to have their bags searched at the city gates for contraband.

Selling these obscene pamphlets was a kind of drug dealing of the day. People had contacts, sources, dealers they would meet on a street corner or in a tavern. When asked by a trusted customer, legitimate booksellers would pull down the shades, lock the door, and remove a floorboard to bring out their stock. Waiters at Paris's 2,800 cafes and taverns rented out the pamphlets for a few hours to clients who copied down or memorized the juicy parts, returned them, and recited the bawdy verses to their fellow patrons amid cheers and applause.

When Marie finally became pregnant in 1778, the *libelles* had a field day speculating on who the father was. Was it her charming brother-in-law? Any one of a number of noblemen at court? A servant? A palace guard? Had she had sex with so many men she couldn't even be sure?

Anti-monarchy pamphlets portrayed the queen as pathologically horny. Their titles included: *The Royal Dildo*, *The Uterine Furors of Marie Antoinette*, and *The Patriotic Bordello*. According to these pamphlets, the queen had first been introduced to sex by her father—who died when she was ten—and then had sex with her older brother. At Versailles, she indulged in drunken lesbian orgies. She kidnapped hapless strangers from the park in Versailles at night and raped them in her carriage. She had threesomes, foursomes, and fivesomes.

One popular pamphlet had her say, "In Olympus, in Hades, I

want to fuck everywhere!" and she does. In another one, she cries, "And if all the cocks that have been in my cunt were put end to end, they would stretch all the way from Paris to Versailles." Another pamphlet claimed that "three quarters of the officers of the Gardes Françaises had penetrated the Queen." In yet another, she said, "Our three interlocked bodies composed the most rare and interesting combinations. Debilitated by our pleasures, exhausted with fatigue, we took time out only in order to mock the misery of the people and drink deeply in the chalice of crime. The brew that filled it served as an omen that, following the example of Caligula, we soon would drink the blood of the French people out of their own skulls." (Yes, that sounds exactly like what someone would say during an orgy.)

Some publishers, realizing how lucrative the market had become, graduated from printing pamphlets to full-fledged books. One bestseller printed in London in two volumes, *The Memoirs of Antonina*, portrayed the queen as preferring a lover like a grenadier "who abridges preliminaries and hastens to the conclusion." *The Libertine and Scandalous Private Life of Marie Antoinette* was published in three volumes, with thirty-two illustrations, and showed Marie, her huge skirts hoisted up, fondling her private parts, or having them fondled by a lady-in-waiting or a nobleman. An illustration in *The Austrian Woman on the Rampage* showed her, legs up in the air and spread wide, on the back of her sleeping husband, while she has sex with his brother. Another illustration showed poor Louis, his limp penis being examined by a doctor who declares it useless.

When her son was born in 1781, one enterprising metal worker created a spoof of the official medal commemorating the blessed event, with Marie cradling her baby on one side, Louis XVI on the other wearing a cuckold's horns. The queen continued to ignore the filth directed at her, while police in France and the French ambassador in London tried to buy up all the copies and destroy them, which only resulted in more being printed and sold.

Oddly, the pamphlets entirely missed out on the one love affair the queen most likely did have, with the dashing Swedish count Axel von Fersen. The count had visited Versailles over the years and flirted with the queen. But something changed in July 1783 when he wrote his sister of how rapturously happy he was to be involved with an unattainable woman. He may have been the father of Marie's second son, born in March 1785, exactly nine months after Fersen visited Versailles. Fersen would, in the coming years, risk his life and devote his fortune to assist the queen, and, after her execution, mark the date every year as a day of mourning. Poor Louis was ridiculed, yes, but for being fat, not terribly bright, and rather useless at sex. Just as the Romans found it unappetizing to take down Mark Antony, the French revolutionaries didn't seem to have the heart to lambaste a man too energetically. The shrill vilification of the monarchy was reserved for Marie. Between the fall of the Bastille in 1789 and her death four years later, the queen found herself in a flood of vile misogynistic slander without parallel since Cleopatra.

The most horrifying accusation came during her trial for treason in October 1793. Marie's eight-year-old son, Louis Charles, had been taken away from her and sent to live with a crude shoemaker who got him drunk, beat him, and made him say his mother had sexually molested him, had lain naked alongside him, with his aunt on his other side, and encouraged him to masturbate. The reason? The queen had done this, supposedly, to have control over him when he became king. Sitting in the courtroom, Marie was speechless when the charge was read. Finally, she said, "If I have not replied it is because Nature itself refuses to answer such a charge laid against a mother." Turning to the crowd, she asked, "I appeal to all mothers here present—is it true?"

"A SECRET SEX FREAK"

Hillary Clinton, too, has been accused of that most sexually depraved crime: pedophilia. In March 2016, an Internet-driven conspiracy theory stated that she and high-level Democratic Party officials ran a child sex ring in the basement of the Comet Ping Pong pizza parlor in Washington, DC (which, incidentally, does not have a basement). Known as Pizzagate, the fiction asserted that kidnapped children were imprisoned in secret rooms where Clinton and her colleagues sexually abused, tortured, and sacrificed them to Satan. One well-meaning but deluded man drove from North Carolina to the pizza parlor where he fired a rifle inside to break the lock on a storage room door (since there was no basement) to liberate all those sexually abused children.

Clinton has also been accused of run-of-the-mill sexual misbehavior. Her opponents have decried her as both frigid and adulterous with men and women, though one must wonder how she could be simultaneously both. In October 2016, Donald Trump said, "Hillary Clinton's only loyalty is to her financial contributors and to herself. I don't even think she's loyal to Bill, if you want to know the truth. And really, folks, really, why should she be? Right? Why should she be?" (Though, come to think of it, if we're curious enough to ask Clinton why she stayed with such an embarrassing philanderer, we might also want to put the same question to Mrs. Trump.)

In the run-up to the 2016 election, the *National Enquirer*, whose publisher David Pecker was an old friend of Donald Trump's, published a WORLD EXCLUSIVE 9-PAGE SPECIAL INVESTIGATION into Hillary Clinton's sex life. "[Her fixer] arranged a lesbian romp for bi-sexual Hillary with a prominent Hollywood identity!" the story gushed. "[He] squealed about a lusty rendez-vous he arranged for Hillary that FINALLY proves the lesbian rumors that have dogged her for decades!" Clinton was an evil nymphomaniac—a "secret sex freak"—who had sex with just about any willing adult.

Numerous heterosexual female politicians have been accused of being lesbians (as if that were a crime): Geraldine Ferraro, the first woman on a major party ticket as Walter Mondale's vice president in 1984, and Ann Richards, who became governor of Texas in 1991. Political opponents of Prime Minister Julia Gillard of Australia accused her of being gay because her partner, Tim Mathieson, was a hairdresser, their public embraces seemed fake and awkward, and their relationship began in 2006, just when her political career was hitting its stride on the national level.

Helen Clark, prime minister of New Zealand, had a deep voice, kept her hair suspiciously short, and had no children with her husband. A 2002 newspaper article reported on "a whispering campaign that she was a lesbian." In an election TV presentation, a TV commentator named Paul Holmes inquired about Clark's intimacy with her husband. He wanted to know if they ever touched each other and accused them of having an "ambiguous marriage." In 2005, an audience member at a televised political debate called her a "no-kids lesbo."

Tarja Halonen, president of Finland, was suspected of being gay due to *her* short hair and her support since the 1980s of gay rights as a matter of social justice. She has said she is heterosexual and has been married to her husband since 2000. A member of the Finnish Parliament, Tony Halme, actually called her a lesbian on a radio interview.

Ironically, in recent years openly gay female politicians have received little criticism for their sexual orientation. Jóhanna Sigurðardóttir, prime minister of Iceland from 2009 to 2013, had been married to her wife since 2002. Ana Brnabić became prime minister of Serbia, a highly conservative country, in 2017, and her partner gave birth to their son in 2019. In the US, Senators Tammy Baldwin of Wisconsin and Kyrsten Sinema of Arizona are gay, as are Representatives Angie Craig of Minnesota and Sharice Davids of Kansas. Though there will always be those who lash out at the supposed depravity of gay people (such as the PAC called Restore American Freedom and Liberty that labeled Baldwin a "pervert" who wanted to teach gay sex to five-year-olds), these women have sustained surprisingly little gay-bashing. Somehow, it's more popular to call heterosexual women lesbians, as if revealing a dirty little secret.

"A HIGH-END CALL GIRL"

Women in elected office all over the world suffer sexually charged slander. In a recent global survey of female politicians, 41.8 percent of them reported seeing embarrassing, sexualized images of themselves on social media.

During an International Women's Day broadcast in 2021, Christine Lagarde, the first woman to head up the European Central Bank, recalled a particularly disgusting comment she was asked as France's trade minister. "And I remember walking to the podium where you have to answer the question and one senior member of parliament looking down on me and saying: 'I wonder who she's gone to bed with to be appointed to where she is,'" Lagarde said. "So I went to see him afterwards to explain to him that I didn't have to sleep with anybody to be where I was, and if he wanted to check my competence and skills he was more than welcome to ask me technical questions that I would be very happy to answer."

Lagarde went on to become the first female finance minister of France and the first female managing director of the International Monetary Fund, replacing Dominique Strauss-Kahn, who resigned after a New York City hotel housekeeper accused him of raping her in his room. In 2020, *Forbes* magazine ranked Lagarde the second most powerful woman in the world.

There seem to be no limits on the implausibility of sexual allegations against women who wield power. On January 21, 2021, Fox News host Mark Levin went on a vicious tirade about eighty-year-old Speaker of the House Nancy Pelosi, which included the accusation that she wanted to have sex with Donald Trump. He said, "Nancy Pelosi, who is a nasty old bag—that's what she is, a nasty, vicious, unhinged fool—she is focused on Trump. She can't get Trump out of her head. I'm starting to think she has—well, let me put it to you this way—an affinity for Trump. May I put it that way, Mr. Producer? The hots for Trump, can I say that, is that legal? She can't get him out of her mind. She can't stop. Maybe her husband can do some kind of—but no, even he can't intercede. Nobody can stop her."

Naturally, as soon as Joe Biden announced Kamala Harris as his running mate, sexually salacious memes and tweets lit up social media portraying her as a wild party girl willing to tumble into bed with almost anybody. These depictions are a far cry from the woman who for years was unusually circumspect in her private life, often reluctant to show up at an event with a date because of the gossip and speculation it would cause, given her high-profile elected position.

What was behind these attacks—just routine, humdrum sexism? After all, Harris has enjoyed an apparently happy marriage to entertainment attorney Douglas Emhoff since 2014. No, there *was* something shocking in her past. In 1994 and 1995, Harris had dated a man still legally married but estranged from his wife for thirteen years, the powerful speaker of the California Assembly, Willie Brown. Brown appointed her to two state boards for which she was well paid, which may have been ethically questionable. But their relationship had been over for about eight years by the time Harris ran for her first elective office.

Once Biden's selection of Harris was announced, Teaparty.org reported, "Flashback: Kamala Harris launched her political career in bedroom as mistress of married mayor Willie Brown." As one meme put it, along with a photo of a smiling Harris shaking Biden's hand, "Pick me as your Vice President! I can do for you what I did for Willie Brown!"

A meme of Harris embracing a pajama-clad Biden in bed reflected her supposed sluttiness and his age. "Pee Pads and Knee Pads," the text ran, "Biden Harris 2020." One meme that circulated widely referred to her as "a high-end call girl."

After the vice presidential debate in October 2020, Eric Trump "liked" a tweet calling Biden's selection of Harris "whorendus."

We can only wonder whether those so morally outraged at a consensual relationship twenty-five years ago with a man separated from his wife would feel kindlier toward her if she had been married three times, cheated on all of her spouses, played the Peeping Tom on countless occasions with fifteen-year-old pageant contestants, had multiple allegations of sexual assault and rape against her, bragged about it, and paid off a porn star to keep quiet about their one-night stand.

And let's face it, speaking of boys being boys, who can forget Bill Clinton's sex with an intern, and New York congressman Anthony Weiner's tweeting photos of his private parts to an underaged girl, and New York governor Eliot Spitzer's 2008 trysts with prostitutes? Does anyone really think New York governor Kathy Hochul will grope, abuse, and intimidate her employees the way Andrew Cuomo did?

No. Because women behave better than that, and yet they are the ones who have been tarred, repeatedly, for thousands of years, with accusations of bad sexual behavior.

CHAPTER 11

SHE'S A MURDERER

Hillary Clinton has personally murdered children. I just
can't hold back the truth anymore.

—Alex Jones, *InfoWars*

While the Misogynist's Handbook paints countless women as whores, there are a few powerful females it has turned into murderers—the worst possible sin, up there with pedophilia—to completely delegitimize them, portraying them as creatures to be reviled and spat upon, silencing their voices forever.

Of course, some female rulers really were murderers. In centuries past, most monarchs of either gender had to execute enemies to ensure their own survival. When Mark Antony had Cleopatra's sister Arsinoe dragged out of a temple and killed, it was probably with the Egyptian queen's approval. Arsinoe had conspired in the past to take the throne for herself.

Cleopatra's younger brother and co-ruler Ptolemy XIV died young right around the time Cleopatra bore Caesar's son and named the infant her new co-ruler. Rumor had it she poisoned her brother, though no one knows for sure. Even if she was behind these deaths, we must consider her actions in the context of her time, her place, and her family. The Ptolemies—descendants of Alexander the Great's general Ptolemy who took over Egypt after Alexander's death—were a murderous lot going back more than two centuries before Cleopatra. They routinely slaughtered mothers, brothers, sisters, and their own children. Such time-honored family traditions were the means of both wreaking revenge and staying alive.

For instance, Berenice II (ca. 267–221 BCE) murdered her husband for sleeping with her mother. When the wife of Ptolemy VIII (ca. 182–116 BCE) tried to replace him on the throne with their twelve-year-old son, the king had his son murdered, dismembered, and sent to the boy's mother as a birthday present. Even if Cleopatra deftly took out her brother and sister, her executions were quite judicious compared to those of her relatives.

When Catherine the Great's husband, the mentally ill Peter III of Russia, died mysteriously of "hemorrhoidal colic" in a cell, it is likely that her lover Grigory Orlov killed him with the empress's consent. Peter had been planning to murder Catherine when she staged a coup and had him arrested. Alive, he would have been a constant focus of rebellious discontent. Catherine's predecessors were truly bloodthirsty: Ivan the Terrible killed and tortured thousands and personally killed his son, while Peter the Great executed thousands more and watched his son be tortured to death. In comparison, Catherine was a merciful and enlightened monarch.

When, in the course of history, queens have executed enemies for crimes, these deaths are remembered with infamy compared to the bloody doings of kings. In the sixteenth century, Mary I of England burned some three hundred Protestants as heretics during her five-year reign, earning herself the sobriquet "Bloody Mary." Yet her father Henry VIII executed some seventy thousand people during his thirty-eight-year reign for a variety of offenses, mostly treason and heresy. No one ever called him "Bloody Henry."

According to the Misogynist's Handbook, most female murderers are sneaky and stealthy (which goes hand in hand with being untrustworthy and inauthentic). Accordingly, in previous centuries, those powerful women believed to be murderers were usually said to have poisoned their victims, slyly slipping a little something deadly into a glass of wine, then serving it with a dazzling smile, whereas men would take the honest route and boldly stab their opponents with a manly thrust.

It is quite true that during the Renaissance, the poison trade was a thriving business—100 percent run by men. In Florence, Italy, for example, the ruling family, the de Medicis, set up a poison factory producing toxins as well as antidotes and testing them on animals and condemned prisoners. Duke Cosimo de Medici attempted to poison Piero Strozzi, a political opponent, in 1548, according to a document in the Medici Archives. An anonymous would-be assassin wrote the duke in cipher, "Piero Strozzi usually stops to drink a few times during his journey." The writer requested "something that could poison his water or wine, with instructions on how to mix it."

In 1590, Cosimo's son Grand Duke Ferdinando, suspected of having poisoned his older brother Francesco to gain the throne three years earlier, wrote his agent in Milan, "You are being sent a bit of poison, and the messenger will tell you how to use it. . . . And we are pleased to promise three thousand scudi and even four to the one who administers the poison. The quantity being sent is enough to poison an entire pitcher of wine, has neither odor nor taste, and works very powerfully. You need to mix it well with wine, and if you want to poison only one glass of wine at a time, you need to take a half ounce of the material, rather more than less."

The mysterious Council of Ten—one of the main governing bodies of the Republic of Venice from 1310 to 1797—ordered assassination by "secret, careful, and dexterous means," a clear reference to poison. In a recent study, Matthew Lubin of Duke University and the University of North Carolina at Chapel Hill identified thirty-four cases of Venetian state-sponsored political poisonings between 1431 and 1767. In all probability, there were many more Venetian poison attempts on political undesirables than were recorded. All done by men. The council hired botanists (men) at the nearby University of Padua to create the poisons.

And yet it is a woman whose name is practically synonymous with poison. Even in her lifetime, Lucrezia Borgia, the illegitimate daughter of Pope Alexander VI, became a cartoon villainess, a woman who reportedly slept with pretty much anyone, including her father and her brother, a devious hussy with a hinged locket ring that she opened while pouring wine, letting powdery white arsenic flutter into the goblet. Highly intelligent and astonishingly beautiful, with golden hair cascading to her knees, Lucrezia had political importance in her own right. In 1499, when she was nineteen, her father named her governor of the prosperous city of Spoleto, a post usually given to a cardinal. In 1501, she was granted official power to run the Church and the Papal States when the pope toured lands conquered by her brother.

The notorious Borgia family had many enemies—which is certainly understandable, given the nepotism, bribery, corruption, murders, and ruthless ambition of the men. One highly effective tool in the toolbox to take down enemy men was to tarnish the reputations of their women. There is absolutely no evidence Lucrezia ever committed incest or poisoned anyone. The poor woman was a pawn moved about on the blood-stained chessboard of her brutal male relatives to advance their own political goals.

Through her third marriage, Lucrezia finally escaped her interfering family (her father had annulled her first marriage, and her brother had murdered her second husband) by ending up in the city-state of Ferrara as wife of Alfonso d'Este, who would become the ruling duke. The d'Este family had not wanted the marriage, fearful of her reputation as a sex fiend and a poisoner, but they had finally caved in to the intimidation and generous financial offers of the pope. But once installed as duchess, Lucrezia made herself universally popular, serving as patron to artists and poets, and reigning graciously over a colorful Renaissance court. She devoted herself to pious works and helping the poor, and capably administered the realm when her husband was away, doing much to reverse her sinister reputation.

But the image of a beautiful blond poisoning incestuous whore is just too good to give up and made its way into the arts. In 1833, the novelist Victor Hugo wrote a play—which soon became an opera—about her poisoning the lovers she tired of. In 1839, Alexandre Dumas wrote a novel featuring her supposed crimes. In the early 1860s, English artist Dante Gabriel Rossetti painted two portraits of her. In one, she is sprawled luxuriously on a chair, her father on one side, her brother on the other, both apparently sniffing her hair. In another, glassy-eyed, she washes her hands, à la Pontius Pilate, after having just poisoned her second husband (who was actually stabbed and strangled on the orders of her brother, but let's just blame the woman). And so Lucrezia Borgia became a nymphomaniac murderer trapped in amber for all time, beautiful, ageless, and treacherous.

"SHE MAY PERHAPS GIVE HER TOO MUCH DINNER ON SOME OCCASION"

Anne Boleyn was accused of every high-level death at Henry VIII's court: the king's powerful minister, Thomas Wolsey, archbishop of York, who died of a nasty bout of diarrhea; Henry's first wife, Catherine of Aragon, who most likely died of cancer of the heart; and Lord Chancellor Thomas More and Bishop John Fisher, both of whom Henry beheaded. The story is that she had Wolsey and Catherine poisoned and forced a weak-willed Henry VIII to execute the others. Bear in mind, Henry had already executed several innocent but inconvenient people by that point. So it's hard to believe that without Anne's insistence the king would have merely sighed with disappointment when More and Fisher steadfastly refused to recognize him as Supreme Head of the Church of England.

True, Anne wasn't fond of Wolsey. In 1522, he was involved in preventing her from marrying Henry Percy, the son of the earl of Northumberland, whom she loved deeply. For dynastic reasons,

Percy was already betrothed to an heiress whom he was forced to marry. Then, over a period of several years, Wolsey had proven ineffective at persuading the pope to dissolve Henry's marriage to Catherine of Aragon. As a cardinal of the Catholic Church, Wolsey could not have been thrilled that the king was intent on marrying an outspoken religious reformer like Anne. It seems Wolsey was intentionally dragging his feet. Finally, Henry had had enough.

In 1529, the king had Wolsey arrested in Yorkshire for high treason. Summoned to London, the cardinal stayed two weeks with the Talbot family at Sheffield Manor Lodge, where he became violently ill with diarrhea, one symptom of poisoning. A witness reported that Wolsey "took to the stool all night . . . unto the next day, he had above fifty stools," and, wandering into the realm of indubitable TMI, added that "the matter that he voided was wondrous black." His captors forced him back on the road, hoping to bring him to the king alive, but he died en route. According to rumor, Anne had had Wolsey poisoned for standing in the way of her ambition to become queen. But a modern examination of his death reveals that his bedroom at Sheffield Manor Lodge was directly above a room filled with human waste, which was shoveled out periodically. It is likely he developed a bacterial infection.

The Spanish ambassador Eustace Chapuys, no friend of Anne's, wrote his master, Charles V, that Anne also wanted to poison Princess Mary, the teenaged daughter of Henry and Catherine. "Indeed," he wrote, "I hear she has lately boasted that she will make of the Princess a maid of honor in her Royal household, that she may perhaps give her too much dinner on some occasion. . . ." In another letter, he informed Charles, "A gentleman told me yesterday that the earl of Northumberland told him that he knew for certain that [Anne] had determined to poison the Princess."

In 1534, Chapuys insisted that Anne was plotting to poison both Catherine and Mary. "Nobody doubts here that one of these days some treacherous act will befall [Catherine] . . ." he wrote. "The King's mistress [Chapuys never recognized Anne as Henry's wife] has been heard to say that she will never rest until she has had her put out of the way. . . . These are, indeed, monstrous things, and not easily to be believed, and yet such is the King's obstinacy, and the wickedness of this accursed woman that everything may be apprehended."

When Catherine died in January 1536 after a long illness, her physician believed she might have been slowly poisoned, even though her food was carefully watched and prepared by faithful servants and tasted by her ladies. Her autopsy revealed a strange black growth on her heart, a sure sign of poison, many believed— probably administered at the direction of that evil femme fatale, Anne. But modern medical experts believe the growth was a cancerous tumor; poison would have affected Catherine's digestive system, not her heart. And if Anne was going to send herself to hell by committing murder, as she surely would have believed, why would she have done it three years into her queenship? Why not during the seven years of Henry's divorce?

SHE "DOTH BATHE HERSELF IN YOUR BLOOD"

When it came to poison, Catherine de Medici, the powerful queen mother of France, had a reputation as notorious as that of Lucrezia

Borgia. This reputation was built upon misogyny and xenophobia—
she was a foreign woman who became quite powerful—as well as
the book *A Mervaylous Discourse*, mentioned earlier, that shredded
her reputation forever and carved in stone the legend of the Sinister Queen.

After the Saint Bartholomew's Day Massacre in 1572, in
which Catholics slaughtered tens of thousands of Huguenots
across France, thousands of survivors left France for the safety of
friendlier nations, many of them landing in that bastion of mili-
tant Protestantism, Geneva. Many of these were well-educated and
prominent individuals who set about creating a concerted public
relations campaign to discredit the French monarchy. And it was
easier to make the woman a villain because, as the *Discourse* states,
women are unfit to rule. A woman is "always young in spirit and has
a will subject to sudden change," whereas a monarch must be calm
and focused. Women "chat and babble" and possess "intemperance
of spirit . . . and unrestrained greed."

And this particular woman, according to the *Discourse*, "layeth
a thousand ambushes, she appointeth a thousand murders." She
"doth bathe herself in your blood" and "delighteth in nothing but
ruin and desolation." The blame-the-evil-woman *Discourse* was a
bestseller, much better than various other books blaming the king
and other important men for the massacre. No less than ten editions were published the first year, in Latin, French, German, and
English, and it continued to be published for decades. Subsequent
histories of the massacre have drawn from the *Discourse*, so that the
story has come down to us almost unquestioned.

The book stated that Catherine's parents were alarmed that the astrologers who cast her horoscope predicted she "should be occasion of great calamities, and of the final and utter subversion of her family or household, also of the place whereunto she should be married." Her parents, however, were in no position to consult with astrologers about their daughter's future. Her mother died of puerperal fever soon after the birth, and her father, who had been languishing in pain for some time, died days after his wife, probably of syphilis.

In 1533, at the age of fourteen, Catherine married the fourteen-year-old second son of King François I of France. François had made the marriage alliance with the ruling house of Florence—who came from a family of bankers—because Catherine's uncle was Pope Clement VII. He was holding out for a more royal bride for his firstborn son and heir, François, duke of Brittany.

According to the *Discourse*, Catherine's first murder occurred when she was only seventeen. In 1536, her eighteen-year-old brother-in-law played a vigorous game of tennis, called for cold water, and was handed a pitcher by Count Sebastiano de Montecuccoli, an Italian courtier who served as the prince's private secretary. Soon after drinking the water, François fell ill of a high fever and died a few days later. Even though the autopsy revealed abnormalities in François's lungs, the king was convinced the Italian had murdered his son and had him pulled apart by four horses. Whatever killed the dauphin, it wasn't poison, which we now know would not have caused his high fever.

But poison it was believed to have been. And who benefited from the prince's death? Catherine, who would be queen of France once the king died rather than the wife of a second son. She, aided and abetted by her murderous Florentine relatives, had arranged the poison through an Italian courtier. Everyone knew about the state-sponsored poison factories in Florence and Venice. When a royal personage in a northern European court died suddenly, heads swiveled to stare at the nearest Italian in the room. A new term arose in England in the sixteenth century; when someone was believed poisoned, it was said he had been "Italianated."

In a 1614 collection of his sermons called *The Devil's Banquet*, the English clergyman Thomas Adams argued, "It is observed, that there are sinnes adherent to Nationes, proper, peculiar, genuine, as their flesh cleaveth to their bones. . . . If we should gather Sinnes to their particular Centers, wee would appoint Poysoning to Italie."

The *Discourse* has Catherine attempting to do away with Admiral Gaspard de Coligny and his brother, François de Coligny d'Andelot, both prominent Huguenots, in 1569. The men were "poisoned at a banquet, whereof the one died and the other very extremely sick, did hardly recover." The fact is that d'Andelot died of a fever at the age of forty-eight—fever, again, indicating natural illness. Catherine is off the hook for this one, too.

The same source accuses her of trying to kill the leader of the Huguenots, Louis de Bourbon, prince of Condé, with a luscious poisoned apple prepared by Master René Bianco, a glover and perfumer who had traveled with Catherine forty years earlier from their native Florence to France. A de Medici retainer who cooked up recipes in his basement laboratory was doomed to have the reputation of a poisoner.

"First therefore to dispatch the prince of Condé," the *Dis-*

course states, "she causeth him to be presented with an apple im-poisoned by a milliner named Master René her perfumer." But the prince's surgeon, La Gross, suspecting "it by reason of the place from whence it came, plucked it out of his hand and smelled unto, whereby presently was procured an exceeding swelling in his whole face." The surgeon then cut it up and gave it to a dog, who died, proving the fruit had been poisoned.

Poison, for Catherine, was "but a sport." She was also blamed for the 1571 death of Odet de Coligny—brother of the admiral and a former Catholic cardinal who had become a Huguenot, mar-ried his mistress, and fled to England. Coligny had been ill for two months when he died at an inn in Canterbury on his return trip to France to join the Huguenot army. His wife was convinced he had been slowly poisoned. The resulting autopsy revealed "the liver and the lungs corrupted," pointing to a natural illness. But they also found spots on the stomach, a perforation of the stomach walls, and lacerated tissues. The chief physician told Coligny's wife that the symptoms were the result of a corrosive agent that ate into the stomach. But in the twentieth century, physicians studying the autopsy report believed Coligny had a painful gastric ulcer (which explained his two-month illness) that suddenly ruptured, allowing his stomach contents to flood his abdomen, and resulting in death within hours.

Poisoned witch-queen apples notwithstanding, throughout the 1560s, Catherine worked tirelessly for Catholics and Huguenots to live in peace in the kingdom. In January 1562, under the Edict of Saint-Germain, she gave the Huguenots limited rights to wor-ship as they wished at prescribed times and in certain jurisdictions.

Huguenots were furious they didn't have full rights to worship any-where and anytime they wanted. Catholics were outraged she had given them any rights at all. When the edict reached the Paris Par-lement for ratification, its members stated that they would rather die than register it.

A priest attached to the court, Claude Haton, wrote in his memoirs, "She went at once to Paris and came close to riding into the palace, horse and all" to show she meant business in getting the edict registered. "Even when she entered the room, her anger had not yet cooled," he recalled, ". . . and she began to plead and weep just as women do when they are angry." She stated that the edict was to save France from plunging into civil war, that she was a good Catholic, but that Huguenots should be treated with greater kindness for the sake of the nation. Still, the Parlement refused, supported by the royal council, the clergy, and the Sorbonne. Cath-erine forced them to register the edict in March, but by then the country was indeed plunging into civil war.

By 1572, after a decade of war, the French economy was devas-tated. Commerce had slowed to a halt. Villages were burned to the ground. Flood and famines compounded the damage. The crown was nearly bankrupt. Catherine, who always saw failure as a tempo-rary condition, decided a royal marriage might bring the two sides together. She insisted that her Catholic daughter, nineteen-year-old Marguerite, marry eighteen-year-old Henri, the son of Jeanne d'Albret, the Huguenot leader and queen of Navarre. Jeanne, prim and puritanical, was horrified at the idea of sending her son to Sodom, as she considered the glittering French royal court, but was finally forced to agree.

On June 4, 1572, the two women, who heartily loathed each other, went shopping in anticipation of the multiday wedding festivities for gowns, jewels, ruffs, gloves, perfumes, and cosmetics at Paris's most popular shops. Though Jeanne dressed severely in black and white with little ornamentation, she had a well-known weakness for beautiful gloves of soft buttery leather, heavily scented with cloves, musk, ambergris, or orange blossom. Queen Catherine took Jeanne to buy gloves at Master René's trendy boutique in the heart of the Paris shopping district.

Jeanne bought a pair of gloves from the perfumer. She must have tried them on before purchasing them and sniffed their scent. Perhaps she wore them home. By the time her carriage rolled up to her lodgings, she felt unwell and went to bed with a slight fever. The following morning, she had a sharp pain in the upper right side of her chest. The physicians were called but could offer her no relief. By June 6, she had difficulty breathing. She died three days later at the age of forty-three.

Evidently, Catherine had killed again. This time with poison-drenched gloves. Never mind the fact that Jeanne had suffered bouts of tuberculosis since childhood and her autopsy revealed a huge leaking abscess on her lung.

Laying down his bloody knife, the royal physician, Desnoeds, said, "Messieurs, if her majesty had died, as it has been wrongly alleged, from having smelled some poisoned object, the marks would be perceptible on the coating of the brain, but on the contrary, the brain is as healthful and free from injury as possible. If her majesty had died from swallowing poison, traces of such would have been visible in the stomach. We can discover nothing of the kind. There is no other cause, therefore, for her majesty's decease, but the rupture of an abscess on the lungs."

But the queen mother murdering another queen with poisoned gloves was just too good a story compared to a boring ruptured abscess. (Nor did they know as we now do that no one could die from touching or smelling a poisoned object given the toxins available at the time.) Poison it was.

With Jeanne gone, the new leader of the Huguenots was Admiral Gaspard de Coligny, who came to Paris with hundreds of Huguenots for the August 18 wedding. Four days later, while walking in the street with his entourage from a palace meeting to his lodgings, the admiral stooped to adjust his shoe. Shots rang out and bullets struck his right hand and his left arm. If he had not bent over, the admiral would have been killed.

Catherine and her son, twenty-two-year-old King Charles IX, went to visit him and sent the palace doctor to attend him. But they were clearly worried. The shot had come from an empty upstairs apartment owned by the de Guise family, whose leader, Henri, believed Coligny had killed his father years earlier. Yet the Huguenots were furious at the French royal family, believing the royal wedding was a stratagem to bring Coligny to Paris to be killed. And, indeed, the *Discourse* has Catherine hiring an assassin named Maurevert to shoot Coligny in the street.

According to contemporary reports, many Catholic leaders feared that Coligny's wounding was more dangerous than his death would have been. Now he was seen as a resurrected religious martyr, the innocent victim of an evil Catholic assassination plot. The admiral, well on his way to recovery, would be more popular and powerful than ever before.

It is difficult to imagine that Catherine plotted the assassination of Coligny, especially after her years of tirelessly playing the peace-maker and just having pulled off the wedding. But it was, alas, to be a red wedding. On the night of August 23, a group of men, led by the duc de Guise, dragged the admiral from his bed, murdered him, and threw him out of the window. They then set upon his followers. People in the street, seeing the most powerful Huguenots murdered before their eyes, followed suit, killing their Huguenot neighbors. The targeted murders of several Huguenots had lit the powder keg of Catholic religious fervor across the nation, as Catholics assumed they had the royal green light to kill the heretics who had caused three civil wars in a dozen years. The slaughter went on for days. Certainly, no one in the royal family expected the assassinations to launch a massacre of, by some estimates, tens of thousands of Frenchmen. The streets of Paris ran red with blood. Entire families—women, grandparents, babies—had been stabbed and lay in heaps. Thousands of bodies were dragged to the Seine River and thrown in.

Naturally, as word got out to stunned monarchs around Europe of the Saint Bartholomew's Day Massacre, suspicion fell on the woman. Instead of blaming the king, who gave the order, it was far more conventional to blame the sinister Italian witch in black—who had already poisoned several people with apples, water, and gloves—for plotting the bloodbath and forcing her weak-willed son to acquiesce. The entire wedding scenario, according to the *Discourse*, was a plot hatched by Catherine for the purpose of murder. Bring all her enemies into Paris for a party, shut the gates, and kill them. The *Discourse* asserted that she had originally tried to arrange the massacre back in 1570 when her son Charles IX married Elisabeth of Austria, but, alas, not enough Huguenots RSVPed that they would be attending the wedding.

Sitting like a big, fat, black spider, Catherine had carefully woven a web of deceit and murder over a period of years to achieve this. And it wasn't just the Huguenot leaders she wanted to kill, according to the *Discourse*; it was pretty much *everybody* with any power, Catholic and Huguenot alike (but, what a disappointment, most of the Catholics got away). Then she could rub her murderous hands together in devilish delight, cackling wickedly as she sat on a throne of grinning human skulls, perched on top of a mound of putrefying corpses.

The *Discourse* has Catherine explaining her desire to massacre everyone of any social standing. "We wish to exterminate all the heads of the nobility," she supposedly said, "those who are born or have become great by notable services, . . . those who could legitimately oppose our evil machinations, those who because of their natural goodness could not assist in our deceits and treacheries." Said no real mass murderer ever.

For more than four centuries, the story of the murderous Catherine—written by an anonymous misogynist Huguenot out to take her down—has become so stuck in popular culture that few have ever questioned it. Yes, she often ruled when her sons proved uninterested in doing so, but when one of them insisted on an action despite her warnings, she would have been incapable of preventing it. Moreover, the massacre went against a solid decade of work—negotiating and temporizing, accommodating and reconciling, doing everything possible to avoid bloodshed and keep the peace. It is extremely unlikely that she would suddenly say, "Oh, fuck it! Let's just kill them all!" and then spend the remaining seventeen years of her life once more laboring ceaselessly for peace. If Catherine were alive today, spending the same energy on stopping hostilities, she would most likely be awarded the Nobel Peace Prize.

In recent years, some historians have reevaluated whether Catherine had anything to do with the massacre at all and are more inclined to pin the blame on her son Charles IX. Unstable and fearful, he believed the assassination attempt on Coligny would cause the Huguenots to rise up in full-fledged civil war again. Better to trap them in Paris and kill them at once—chopping off the head of the snake, so to speak—rather than allowing war to once more devastate the countryside. Perhaps he feared the enraged Huguenots would even try to kill members of the royal family in revenge for the attempt on their leader's life. Eager to prove his power as the weak often are, Charles IX may have seen the assassinations as a show of royal strength.

Other historians point the finger at Henri, duc d'Anjou, the king's younger brother. The French historian Thierry Wanegffelen wrote that in a royal council meeting held soon after Coligny's wounding, Catherine's advisors recommended the assassination of some fifty Huguenot leaders, a suggestion she adamantly opposed. But d'Anjou, who served as Lieutenant-General of the Kingdom, saw the move as making a grand name for himself among Europe's Catholics, increasing his status from that of little-known younger son. It was, according to Wanegffelen, he who convinced the king to proceed, and he who gave the orders to the Paris authorities to close the city gates and chain the boats in the river so no Huguenot leaders could escape.

Charles, a bit unbalanced even before the massacre, became completely unhinged afterward. Sometimes he boasted about how successful his massacre had been. Other times he ran around the palace with his fingers in his ears claiming he could hear the murdered Huguenots screaming. "What blood shed! What murders!" he would shriek. Other times he blamed his advisors. "What evil counsel I have followed!" he would cry. "O my God, forgive me . . . I am lost! I am lost!" Sometimes he blamed his mother. "God's blood, you are the cause of it all!" he shouted at her. Catherine sadly replied that her son had become a raving lunatic.

When he died in 1574 at the age of twenty-three, coughing up blood—a sure sign of tuberculosis—rumor had it that Catherine had killed him by mistake. She had commissioned a poisoned book on falconry to murder her son-in-law Henri of Navarre, its lovingly colored pages sprinkled with arsenic. But Charles touched it first, turning the poisoned pages, then touched his lips, threw up, and died within hours. It's a good story, but such a tiny amount of arsenic would not have killed him. And, young as he was, he had been in a steady physical and mental decline for years; tuberculosis seemed to have been the least of his problems.

The *Discourse* lets the question dangle as to whether she intentionally poisoned her son, as he had been struggling against her stranglehold on power for greater independence. Because killing her own child would not have been too much for such a murderous witch monster, right?

Two centuries later, the French underground press accused another foreign queen, Marie Antoinette, of murdering her firstborn son, who had never been truly healthy since birth and died of tuberculosis at the age of seven in 1789. She also intended to poison her husband, Louis XVI, and make her supposed lover, his younger brother, Charles-Philippe, comte d'Artois (who was not her lover), king, so they could rule together. She had, according to the *libelles*, already poisoned two royal ministers, Maurepas and Vergennes, using an old recipe of Catherine de Medici's. Actually, she wanted to kill everybody just for the malicious fun of it. Engravings depicted her, breasts exposed, with a dildo in one hand, concocting poisons with the other. In a 1789 play called *La Destruction de l'Aristocratisme*, she so despised the French people that she cried, "With what delight I would bathe in their blood!"

THE MANY MURDERS OF HILLARY CLINTON

The Catherine de Medici of our time, Hillary Clinton has been blamed for the deaths of just about everyone she ever knew who did not die of cancer in a hospital. So far, the Clinton hit list numbers some fifty individuals who got in the way of her political ambitions, according to conspiracy theorists, though in many cases, just how exactly the victims did so is unclear, so perhaps, like Catherine de Medici and Marie Antoinette before her, she supposedly killed out of sheer malice. Her many murders—some of them believed to be aided and abetted by her husband—were brought up during both her 2008 and 2016 presidential runs. There is even a hashtag, #ClintonBodyCount, attributed to Linda Thompson, a lawyer and conspiracy theorist at the American Justice Federation, an organization whose main purpose seems to be churning out ridiculous accusations.

The first alleged Clinton murders were those of two Arkansas youths, seventeen-year-old Kevin Ives and his friend sixteen-year-old Don Henry, who had set out in the middle of the night to go hunting and were run over by a train at 4 a.m. on August 23, 1987. Initial findings indicated the boys had passed out on the train tracks from smoking too much marijuana. Their parents insisted on a second autopsy, which showed one had been stabbed and the other's skull had been crushed by a blow from his gun. A week before the deaths, in the same area where the boys were discovered, a police officer had spotted a man in army fatigues who fired on him and disappeared into the woods. One possible reason for the boys' murders was that they had witnessed a drug deal. Somehow that morphed into the Clintons being involved in the theorized drug deal and ordering the boys killed.

Hillary Clinton's most notorious alleged murder was that of Deputy White House Counsel Vincent Foster, a good friend of hers from the Rose Law Firm in Little Rock, Arkansas, where they had both worked. In 1993, Foster was named deputy White House counsel and moved to Washington, DC. A quiet, gentlemanly soul, he quickly felt out of place in the brutal arena of DC politics. The media excoriated him for botching Department of Justice nominations and his handling of a scandal involving the White House travel office. Foster fell into a deep depression, noticed by many concerned family members and friends and diagnosed by a psychiatrist, who prescribed medication. Only six months into the job, Foster shot himself in the mouth in a Virginia park. In a note, found torn to pieces in the bottom of his briefcase after his death, he wrote, "The WSJ [Wall Street Journal] editors lie without consequence. . . . I was not meant for the job or the spotlight of public life in Washington. Here, ruining people is considered sport."

Soon, rumors abounded that Hillary Clinton had, in fact, had him murdered because he knew too much, or was going to reveal their affair, or for some other nefarious reason that she has, in her devilish way, managed to hide.

Conspiracy theorists asserted that there was no exit wound in Foster's head. The gun that killed him had been placed in his hand after his death. On his radio show, Rush Limbaugh crowed, "Vince Foster was murdered in an apartment owned by Hillary Clinton." There wasn't much blood at the suicide scene; proof, it seemed, that he had died elsewhere and the body had been moved. The *New York Post* claimed that investigators "never took a crucial crime-scene photo of Vincent Foster's body before it was moved" out of the park where it had been discovered and started putting quotation marks around the word "suicide."

The tales of the Clinton murder roiled the stock market because, as a highly respected Lehman Brothers analyst explained, traders "were afraid Hillary Clinton was involved in a murder," she said. "They hate that." Yes, it is always bad news for the stock market when the first lady murders her lover.

An ABC News team of investigators, however, saw the gruesome crime scene photos, replete with plenty of blood and powder burns on Foster's hand. Over the course of the next three years, five official investigations concluded that Foster had died by suicide. The first was undertaken by the US Park Police, in whose jurisdiction the body had been found, assisted by the FBI and several other state and federal agencies. Released three weeks after the death, on August 10, 1993, the report stated, "The condition of the scene, the medical examiner's findings and the information gathered clearly indicate that Mr. Foster committed suicide."

In June 1994, Independent Counsel Robert B. Fiske issued a fifty-eight-page report based on the opinions of several pathologists that asserted, "The overwhelming weight of the evidence compels the conclusion . . . that Vincent Foster committed suicide in Fort Marcy Park on July 20, 1993."

In August 1994, Representative William F. Clinger Jr. of Pennsylvania, the ranking Republican on the House Committee on Oversight and Reform, also concluded suicide as the cause of death. In January 1995, the Senate Committee on Banking, Housing, and Urban Affairs agreed with the findings. Finally, after a three-year investigation, Independent Counsel Ken Starr, who also worked on the Whitewater investigation, released a report on October 10, 1997, also affirming that the death was a suicide.

Hillary and Bill Clinton were blamed for the plane crash of C. Victor Raiser II, who served as national finance co-chairman for Bill Clinton's 1992 presidential run. On July 30 of that year, Raiser, his son, and three other passengers crashed while going on a fishing trip to Alaska. The National Transportation Safety Board stated that it was pilot error. Why, we ask, would the Clintons kill Raiser? Well, clearly, he must have known something horrible about them and was threatening to reveal it.

A former White House intern (no, not that intern) named Mary Mohane was killed in 1997 along with two other employees while working at a DC Starbucks during a robbery in which she tried to take the robber's gun. Two years later, the robber was found, confessed, and later convicted. Why would the Clintons kill Mohane? She must have had an affair with Bill, too, and was going to testify for special prosecutor Ken Starr in the Monica Lewinsky investigation.

Hillary and Bill Clinton have also been blamed for the death of a former business partner, James McDougal. The Clintons, James McDougal, and his wife, Susan, had invested in a failed Arkansas real estate venture known as Whitewater. An in-depth investigation of the Clintons for fraudulent activity yielded no evidence, but McDougal was convicted on eighteen felony counts of fraud and conspiracy. In March 1998, McDougal, who had been diagnosed with a heart condition, died of a heart attack in solitary confinement, where he had been sequestered as a penalty for refusing to provide a urine sample for a drug test. The Clintons had somehow killed McDougal in prison, according to conspiracy theories, and made it look like a heart attack because he knew of their guilt and was going to inform the authorities.

In April 1996, Ron Brown, US Secretary of Commerce, died in a plane crash in Croatia while on a trade mission, along with thirty-four others. In the crash, a bolt had punctured his skull. Conspiracy theorists alleged that the puncture had been a bullet hole and that X-rays of Brown's skull revealed bullet fragments. Though how they imagine Brown was shot was never revealed. Did someone shoot him on the plane just before it crashed? Did the shooter parachute off the plane to safety or was he killed in the crash, too? Or did someone shoot Brown, carry his dead and bloody brain-bespattered body on board, and prop it up in his seat, while no one noticed?

Examination of autopsy photos of his skull showed there were no bullet fragments, just the blunt-force trauma of the bolt. And an Air Force investigation found the cause of the crash to be pilot error. Nor could proponents of this fantasy give a plausible explanation as to why the Clintons would want their commerce secretary dead. They merely intimated there was a corrupt business deal involved.

Hillary Clinton has also been blamed for the unsolved 2016 murder of a Democratic National Committee staff member, twenty-seven-year-old Seth Rich. On the night of July 10, 2016, Rich had been to Lou's City Bar, less than two miles from his DC apartment. He left the bar around 1:30 a.m. At 2:05 a.m. he called his girlfriend and spoke for about two hours as he ambled home. At 4:20 a.m., only a block from his apartment, shots rang out. Police found him unconscious with two gunshot wounds in his back and, apparently, nothing stolen. He died ninety minutes later at the hospital.

Within thirty-six hours of the murder, online commenters were spinning wild stories. One Reddit user wrote, "Given his position & timing in politics, I believe Seth Rich was murdered by corrupt politicians for knowing too much information on election fraud." Conspiracy theorists, later aided and abetted by Sean Hannity and Lou Dobbs of Fox News, Alex Jones of *InfoWars*, and former Speaker of the House Newt Gingrich, theorized that Rich had given Julian Assange, founder of WikiLeaks, tens of thousands of DNC emails. The Russians, innocent as newborn babes, hadn't done it; Seth Rich had done it. And Hillary had made him pay.

There were several things wrong with the theory, however. For one thing, the Russians *had* done it, according to numerous US intelligence agencies. It was also contradicted by the July 2018 indictment of twelve Russian military intelligence agents for hacking the email accounts and networks of Democratic Party officials. Rich had not had a high-level position at the DNC, only that of a staffer designing a computer application to help voters find their polling places. His computer skills, according to those who knew him well, were certainly not on the level to do a massive system-wide hack. Moreover, he was shot in the back, not in the head, which would have been typical for a professional hit job. And the neighborhood had seen several robberies that summer. It was, most likely, a robbery gone bad. Rich had bruises on his hands, face, and knees; clearly, he had been trying to fight off his attacker.

Given the number of people Hillary Clinton and her husband are purported to have killed to protect themselves, it is odd that those who truly threatened the couple are still walking the earth: Bill's mistress Gennifer Flowers, who almost derailed his 1992 presidential campaign; Monica Lewinsky for that infamous little blue dress; Ken Starr for his special prosecutor investigation into Bill's perjury about his relationship with Lewinsky; and political opponents Barack Obama, Bernie Sanders, and Donald Trump. Each of them could have easily been taken care of with a plane crash, a train running over them, a bullet, or a shiny poisoned apple.

And last but not least, there are the children Hillary Clinton abused and killed beneath the Comet Ping Pong pizza parlor. In November 2016, Alex Jones recorded a YouTube video, now mercifully offline, in which he stated, "When I think about all the children Hillary Clinton has personally murdered and chopped up and raped, I have zero fear standing up against her. Yeah, you heard me right. Hillary Clinton has personally murdered children. I just can't hold back the truth anymore."

CHAPTER 12

ADDITIONAL TOOLS TO DIMINISH HER

*The history of men's opposition to women's emancipation
is more interesting perhaps than the story of that
emancipation itself.*

—Virginia Woolf, *A Room of One's Own*, 1882–1941

The Misogynist's Handbook offers a variety of microaggressions to diminish and delegitimize powerful women. Several of these have to do with how the woman is identified.

1. MISPRONOUNCE HER NAME

Kamala Harris's first name is pronounced "comma-la," as she writes in her biography, "like the punctuation mark." Harris explained that her name means "lotus flower" and has important symbolism in Indian culture because "a lotus grows underwater, its flower rising above the surface while its roots are planted firmly in the river bottom."

She has also made clear on numerous occasions it rhymes with "Momma-la," which is what her stepchildren call her. Yet no sooner had Joe Biden announced Harris as his running mate than Republican commentators on television deliberately mispronounced her first name. Vice President Mike Pence called her "Kah-MAH-lah," emphasizing the second syllable numerous times during a campaign appearance in Iowa. Republican National Committee Chairwoman Ronna McDaniel did the same. At a September 8, 2020, rally, Trump mispronounced her first name three times in a row, with great exaggeration, as his audience booed. Appearing on Fox News, Trump's personal attorney Rudy Giuliani was, at least, a bit more inventive, calling Harris "Pamela."

Fox News host Tucker Carlson was miffed when a guest corrected him after he pronounced "Kamala" incorrectly. "So what?" he said, before mispronouncing her name yet again, then whining about liberals being too sensitive, and finally ending with that most dismissive of words, "Whatever."

David Perdue, a senator from Georgia from 2015 to 2021, served three years with Harris, working with her on the budget committee. But when he spoke at a Trump rally in Georgia, he called her "Ka-mal-a, Comma-la, Ka-Mala-mala-mala." He, too, ended with a "whatever" as the crowd clapped and cheered.

To be sure, many well-meaning people make honest mistakes with unusual names. At various 2020 campaign events, some supporters who introduced Harris flailed horribly; one even calling her "Camille." Joe Biden himself mispronounced the first "a" in Kamala during his speech introducing her as his running mate, but he quickly corrected himself.

Mispronouncing an individual's name intentionally and repetitively signals the person is not worthy of one of life's most basic courtesies. And for people of foreign or non-white cultures, it can indicate they are un-American, different, difficult—they don't even have normal names, for God's sake—and they should probably go back to the shithole countries from whence they came where everyone can revel in their odd and ungainly appellations.

The issue is not limited to politics. Rita Kohli, an education professor at the University of California, Riverside, coauthored a 2012 study that identified the intentional mispronunciation of a person's name, especially a name tied to a particular culture, as a "racial microaggression." Kohli told the *Washington Post* in October 2020 that the crowd that cheered Perdue's butchered pronunciations of Harris's first name were "cheering the idea that she's not from here, she's not American so we can't take her seriously. There's a deprofessionalization and othering that happens that we wouldn't see them do to Joe Biden."

By mispronouncing the name of Harris—the daughter of a Jamaican father and an Indian mother—political opponents are indicating she is not worthy of the second-highest political position in the nation; she is too foreign and different. In Harris's case, the mispronunciation is a triple-barbed blunder: it is sexist, racist, *and* xenophobic.

"It is an effort to diminish her," Fatima Goss Graves, president of the National Women's Law Center Action Fund, told the *Associated Press* several days after Biden's announcement. "It's designed to signal difference."

In October 2020, Representative Pramila Jayapal of Washington State told the *Washington Post*, "I think it's been happening more and more during the Trump administration. I mean, this is a sitting US senator who he's mocking and who is the first woman of color on a major party ticket—that's not all a coincidence. That's not only planned, but it's the result of a president who has done everything he can to otherize and rile up crowds to do the same."

Jayapal—whose last name is pronounced "JYE-uh-pal" and is also of Indian heritage—has experienced the intentional mispronunciation of her last name on many occasions. In the 2020 election, her Republican opponent, Craig Keller, mispronounced her name at least a half-dozen times during a candidate forum, even after she asked him to pronounce it correctly. When the *Washington Post* inquired as to why he did so, he emailed the paper a bizarre document in which he called her "Jail a pal." He wrote, "Truly! How does one correctly pronounce it! 'Jai a pal', 'Jay a pal' or 'Jail a pal'?"

2. DENY HER HER PROPER TITLE

Another means of diminishing a woman's stature is to deny her the title of respect she has earned. It's an old one. In the eighteenth century, King Frederick the Great of Prussia, who hated the then-current monstrous regiment of women rulers, refused to call Empress Maria Theresa of the Austro-Hungarian empire by that title;

he called her the queen of Hungary, both a much smaller region to reign over and a far less important title. European queens were a dime a dozen; she was one of only two empresses. Frederick shrank her down to what he considered an appropriate size in three words.

In the 2008 presidential primary, both Hillary Clinton and Barack Obama were senators. Yet Clinton's title of "Senator" was omitted 15 percent more than it was for Obama.

In the 1990s, French female cabinet ministers were addressed as Madame le Ministre—Madam the (male) minister, which they found weird. They requested to be addressed as Madame la Ministre—Madam the (female) minister. The Académie Française, which has since its creation in 1634 held a stranglehold over the French language, steadfastly opposed such a shocking change, evidently finding it too jarring to have a feminine article before the word "minister." In 1647, one of its founding members, Claude Favre de Vaugelas, wrote, "The masculine gender is the noblest one. Therefore, it should dominate each time both genders are put together." After decades of pressure, the Académie relented in 2019, agreeing to feminize all professions and titles.

The exact same issue arose in Italy, which also has masculine and feminine articles. The speaker of the parliament is called the president of the Chamber of Deputies. But when Laura Boldrini assumed the position in 2013, her colleagues kept addressing her as the (male) president because "president" is a masculine noun in Italian—*il presidente*. She wanted them to use language appropriate to her gender—*la presidente*—which seemed like a no-brainer to her, but they continued to call her by the masculine form, even though she was the third woman to hold the post.

Boldrini pointed out to *BuzzFeed News* in 2018, "Language is not only a semantic issue, it is a concept, a cultural issue. . . . When you are opposed to saying *la ministra* or *la presidente*, it means that culturally you are not admitting that women can reach top positions. Everything must remain masculine."

When she sent a letter asking her colleagues to use feminine articles for female people, the response was immediate and indignant. She was accused of trying to cancel the Italian language. She was wasting the taxpayer money by having her stationery reprinted. She was insulting the dignity of women. She was waging war on the centuries-old Italian culture and traditions. "In Italy there is a real difficulty in accepting the authoritativeness of women," she said.

Similarly, in 1976, on Maxine Waters's first day in her first elected position as a member of the California State Assembly, she introduced a motion to formally change the title of those officials from "assemblyman" to "assembly member," as the body had some women. Perhaps because she rolled over them with the force of her oratory, stunning them into submission, the dazed assembly approved her motion by a vote of 48 to 27. By the time the assemblymen realized what they had done, however, as if waking from a trance, they reopened the debate and overturned her motion, 41 to 26. "Men attacked me viciously," she told the *Los Angeles Times* in 1992. "They charged me with trying to neuter the male race." (Hmmm, that old castration story again.) The rejection of Waters's motion reinforced the notion that only men belonged in the assembly.

In December 2020, the *Wall Street Journal* published an op-ed by columnist Joseph Epstein, who asked First Lady Jill Biden not to refer to herself as "Dr." since her degree was in education, not medicine. Biden had earned the degree at the age of fifty-five after fifteen years of study while raising three children. In his piece, Epstein called her "Mrs. Biden—Jill—kiddo"—the last one particularly insulting and disrespectful for a sixty-nine-year-old soon-to-be first lady of the United States. "'Dr. Jill Biden' sounds and feels fraudulent," Epstein wrote, "not to say a touch comic." The uppity woman was boasting of a very dubious achievement, Epstein indicated, as "no one should call himself Dr. unless he [*he!*] has delivered a child."

What was at the root of this unmerited attack? On the personal level, probably sour grapes. Epstein, it should be noted, had only a bachelor's degree. And people like Jill Biden—female people, that is—don't deserve to have a higher educational degree than a man. On the larger level, Epstein's attack was a classic tool from the Misogynist's Handbook to put successful women in their place by denying them their accomplishments that are increasingly threatening to an increasingly fragile Patriarchy.

Doug Emhoff, husband of then Vice President-elect Kamala Harris, offered his support to Dr. Biden and all women who suffer the slings and arrows of achievement diminishment. "This story would never have been written about a man," he tweeted. It was certainly never written for Richard Nixon's secretary of state, Dr. Henry Kissinger, whose Harvard PhD was in government. Nor, as far as we know, did Dr. Kissinger ever deliver a child or snip out someone's tonsils.

In a final sexist recommendation reeking of the 1950s, Epstein suggested, "Forget the small thrill of being Dr. Jill." (Excuse me, *small thrill?*) "And settle for the larger thrill of living for the next four years in the best public housing in the world as First Lady Jill Biden."

Yes, she should really be more modest and forget all that accomplishment bullshit. Her husband got her a lovely *free house*. With butlers. Maybe she could redecorate.

3. CALL HER BY HER FIRST NAME

One way to diminish female politicians is to identify them by their first name, whereas men in similar positions are usually identified by their last name. As prime minister of Australia, Julia Gillard found that many journalists couldn't seem to bring themselves to call her prime minister. It even seemed beyond some of them to call her "Gillard." She was "Julia." In 2012, for instance, the *Australian* newspaper had a banner headline: "What Julia Told Her Firm." Journalists rarely, if ever, called her predecessors in the office "Tony," "Kevin," "John," or "Paul." They were called "Abbott," "Rudd," "Howard," and "Keating."

Nicola Sturgeon, first minister of Scotland and leader of the Scottish National Party since 2014, is often called "Nicola" in the media. But her male predecessor, Alex Salmond, was called "Salmond," and his predecessor, Jack McConnell, was called "McConnell."

In 2017, a French politician, conservative economy minister Bruno Le Maire, welcomed two new colleagues in a transfer of power ceremony aired on national television. The male he referred to by his full name: Benjamin Griveaux. The female, Delphine

Gény-Stephann, he referred to as "Delphine" twice. Marlène Schiappa, France's gender equality minister, pointed out the discrepancy. "Calling a female politician by her first name and her male counterpart by his full name amounts to everyday sexism," she said. "It's a bad habit that male politicians need to shake off."

According to a 2007 book called *Rethinking Madam President* by professors Lori Cox Han and Caroline Heldman, "Gendered language of this sort is not consciously disrespectful, perhaps, but gender difference is not random and has the 'real world' consequence of delegitimizing knowledge, experience, and ultimately, leadership." Using only a woman's first name makes her seem like a child. Or a dog. Something adorable and friendly, perhaps—and we all know women need to appear more likable—but not deserving of the respect given an adult.

A 2018 study titled "How gender determines the way we speak about professionals" published in the *Proceedings of the National Academy of Sciences* concluded, "We find that, on average, people are over twice as likely to refer to male professionals by surname than female professionals. Critically, we identified consequences of this gender bias in speaking about professionals. Researchers referred to by surname are judged as more famous and eminent. They are consequently seen as higher status and more deserving of eminence-related benefits and awards."

A 2010 study in the *Political Research Quarterly* called "What's in a Name? Coverage of Senator Hillary Clinton during the 2008 Democratic Primary" found that the media was five times more likely to call Hillary Clinton "Hillary" than they were to call Barack Obama "Barack." The study also found the use of her first name was not to differentiate her from her husband, Bill, the former president. Media could have called her "Hillary Clinton," "Secretary Clinton," or "Senator Clinton," but chose instead to call her "Hillary." Nor was the frequent use of only her first name due to the fact that she campaigned as "Hillary." Journalists referred to male candidates who campaigned under their first names (Jeb! Mayor Pete. Rudy) by their last or full names.

4. DON'T USE HER NAME AT ALL: CALL HER "SHE"

During Julia Gillard's tenure as prime minister, her opponent Tony Abbott constantly called her "she" and "her" in his press appearances. Others followed his lead. For example, on August 21, 2012, during Question Time in the House of Representatives—that raucous free-for-all in the parliamentary system—Christopher Pyne, the manager of opposition business, interrupted Gillard, who was answering a question. "Madam Deputy Speaker," Pyne said, "on a point of order. She is defying your ruling. You asked her to be directly relevant and it was a very specific question."

The leader of the House, Anthony Albanese, interrupted, pointing out "the standing order which requires that people be referred to according to their titles. 'Prime Minister' is the title."

When speaking about Governor Gretchen Whitmer of Michigan, Donald Trump refused to use her title or name, referring to her as "the woman in Michigan."

Even the chant "Lock her up!" uses this misogynistic diminishing tool. In a July 2020 interview with the *Washington Post*, cognitive linguist George Lakoff pointed out that the chants don't use the name of the individual who should be locked up, just the feminine pronoun, which make her "not a person. She's this thing that's out there that should not be paid attention to—that should be gotten rid of."

5. KEEP REPEATING THAT SHE IS A "FEMALE" LEADER

When the words "female" or "woman" are placed before her position (president, vice president, prime minister, senator), the public sees the individual as "not simply a politician (male as norm) but a special kind of deviant professional, a woman politician," according to a 1996 study by researchers Annabelle Sreberny-Mohammadi and Karen Ross. For instance, 55 percent of articles reporting Julia Gillard's leadership challenge made note of the fact that she was a woman. None pointed out that her opponent Kevin Rudd was a man. And how many times have you read the words "male president Joe Biden"? It goes without saying, right?

As Gloria Steinem said, "Whoever has power takes over the noun—and the norm—while the less powerful get an adjective."

"Gender markers reveal the unspoken cultural understanding that politicians, senators and candidates must be men," wrote Dr. Lindsey Meeks in a 2012 study. Because male is the default, and female is this weird alien thing who really doesn't belong.

6. COMPARE HER TO A DOLL

Kim Campbell, Canada's first female prime minister, was sometimes compared to a blond doll in the press. She was called a "straight right-winger with fluffy blond hair" and a "glassy-eyed, tense, blonded doll." Her winning the top job unleashed a torrent of dumb blonde jokes.

In 2017, Canadian MP Gerry Ritz tweeted a link to a news article with the headline "No major advanced industrialized economy is currently on pace to meeting its Paris commitments," adding, "Has anyone told our climate Barbie!"

Catherine McKenna, then the minister of environment and climate change, was the target of the insult, commonly used by her political opponents. She tweeted back, "Do you use that sexist language about your daughter, mother, sister? We need more women in politics. Your sexist comments won't stop us."

When Nancy Pelosi won the post of minority leader of the US House of Representatives in 2002, the conservative talk show host Rush Limbaugh photoshopped her head on a beauty queen's body on his website and labeled her "Miss America." The editor of the right-leaning *Washington Times* called her the party's "new prom queen."

In 2016, the leader of Italy's right-wing anti-immigrant Northern League party, Matteo Salvini, held up a blow-up sex doll at a rally and referred to Laura Boldrini, the president of the Chamber of Deputies. "Boldrini's clone is here on the stage," he said.

"Women are not dolls," Boldrini posted on Facebook, "and the political battle is carried out with arguments—for those who have any—and not with insults." During a televised interview with Salvini, she pointed out, "You realize that's demeaning to women." When he refused to apologize, she held up a sign with the hashtag #WomenNotInflatableDolls.

The media often dubbed Sarah Palin, the 2008 Republican vice presidential candidate, as "Caribou Barbie," a reference to her love of hunting. An enterprising manufacturer produced blow-up sex dolls of her with "bursting cleavage and sexy business suit." Instructions offered the advice to "blow her up and show her how you are going to vote. Let her pound your gavel over and over. . . . This blow up sex doll could really satisfy those swing voters."

British news organizations, in particular, are guilty of trivializing female MPs by referring to them as a kind of harem associated with their male party leaders: "Blair's Babes," "Dave's Darlings," "Cameron's Cuties," "Gordon's Gals," and the truly horrifying "Nick's Nymphets."

7. PROVE THAT SEXIST THING YOU DID IS NOT SEXIST BECAUSE YOU HAVE A WIFE AND DAUGHTERS

Sexists who have been caught being sexist often trot out their wives and daughters and say, "Wait! Look who I live with! People with breasts! Ovaries! You can't accuse me of hating women!"

After Congressman Ted Yoho of Florida called Congresswoman Alexandria Ocasio-Cortez a "fucking bitch" on the Capitol steps, he stood up in the House and trumpeted that he had been married for forty-five years and had two daughters, intimating that there's the proof that he couldn't *possibly* be sexist. (Naturally, the Patriarchy doesn't use the handbook against wives and daughters who are not threatening male power in any way. They are not running for political office or aiming for the position of CEO, thereby usurping places that rightfully belong to men.)

Margie Abbott, responding to Prime Minister Julia Gillard's 2012 charges of sexism against her husband, Australian politician Tony Abbott, gave a speech in which she said, "Don't ever try and tell me that my husband of twenty-four years and father of three daughters is on some anti-woman crusade. It's simply not true." (In an interview published on the same day, she added that he even loves *Downton Abbey*, irrefutable proof he couldn't be sexist, even though he said in an interview, "If it's true that men have more power generally speaking than women, is that a bad thing?")

8. USE GENDERED WORDS TO DESCRIBE HER

Certain words are gender-coded terms for "unacceptable female." These include: feisty, bossy, emotional, shrill, loud, shrill, pushy, angry, shrill, high-maintenance, unlikable, robotic, ambitious (which is a negative for a woman, but not for a man), and shrill. Those who don't like a particular woman should make the effort to find other, non-gender-coded words to describe why, being very specific about which of her words or actions are objectionable. If in doubt, change her name to that of a man and see if you would still say or write the same thing about him.

9. FOCUS ON HER BODY PARTS

Another way to diminish women is to reduce them to body parts. Making fun of the size of Hillary Clinton's ankles, for instance, calling them "cankles." Or Michelle Obama's rear end (which isn't only sexist but is also racist).

Perhaps the strangest obsession with a female politician's body part is with Julia Gillard's earlobes. "So big they could have their own seat in parliament," crowed one commentator. After a televised debate, one blogger wrote, "I can't remember a thing from the debate . . . just those earlobes." Another posted that they were so huge she could wear all her earrings at once.

10. VILIFY HER WHEN SHE MAKES A MISTAKE BUT NOT HIM

In May 2017, British Labour MP Diane Abbott stuttered and stammered in an interview about the cost of police recruitment, finally stating that each new officer would be paid a salary of a few pounds a year. Abbott was immediately derided as incompetent, stupid, fat, and a waste of space. She revealed soon after that she was a diabetic, that the controversial interview had been her seventh of the morning—the prior six had been flawless—and she hadn't had the time to eat. Her blood sugar had plummeted, and her thinking had become foggy.

Two weeks later, the Chancellor of the Exchequer Philip Hammond understated the cost of a high-speed railway by £20 billion pounds. Initially, most major news outlets failed to report it. Two weeks after that, when an interviewer asked MP Andrew Mitchell if he knew what the minimum wage was, he said, "Less than £9." When his interviewer shook her head, he guessed, "About £6." He finally settled on £8. In fact, minimum wage was £7.50 for those over twenty-five. Nor could Mitchell say how many people were on the housing waiting list. While the interview made headlines,

Mitchell did not receive the vitriolic abuse that Abbott did. Forty seconds of a botched interview greatly diminished thirty years of Diane Abbott's service in Parliament as a competent and popular politician. Because women are held to far stricter standards than men.

11. GIVE HER A REALLY LOUSY TOILET IN A GALAXY FAR, FAR AWAY WITH NO TAMPONS

When the first women were elected as British MPs in the 1920s, their office, the Lady Members' Room, was a tiny place in a dingy basement called "the dungeon," which lacked enough desks and chairs for all of them. Some of the MPs sat on the floor, writing letters on their knees. To get to the nearest toilet, the female MPs had to walk down three long corridors and up two staircases. Any lady member feeling the least twinge of her bladder or bowel would have to get a running start to make sure she got there in time. We can picture these poor souls, racing very unladylike through the hallowed halls of Parliament, knocking down any hapless individual who got in their way. New toilet facilities closer to the Lady Members' Room were not constructed until the 1960s.

Similarly, when Pat Schroeder arrived in Congress as a freshly minted thirty-two-year-old Colorado representative in 1973, her initial excitement was tempered by the pathetic toilet facilities. She realized she would need to fight, she wrote, for "a place where we could pee. . . . There were men's bathrooms right off the main floor of the House, but the ladies' room was at the other end of the earth, constructed out of the original Speaker's Lobby in the Old Capitol, and it looked as if it hadn't been updated since the inception of indoor plumbing." She added, "The assumption was that we should be so appreciative of being allowed into the halls of Congress, we'd fall on our knees in gratitude for every crumb."

A new facility for women legislators just off the Senate floor was not made available until 1993, when fifty-four women were members of Congress. But the new restroom was small and windowless. Three stalls were added in 2000. In 2011, a new restroom was constructed off the House floor to accommodate the ninety congresswomen then in office. Since 1962, they had been forced to use the restroom inside the women's reading room. In 2013, the Senate doubled the number of women's restrooms for the one hundred congresswomen serving at the time.

Despite the increased number and more convenient locations of women members' restrooms, they are fairly substandard considering their users are the nation's top leaders. Feminine hygiene dispensers are either nonexistent or often empty. "I have never been in a bathroom that didn't have a machine with feminine products," Congresswoman Norma Torres of California told *Apartment Therapy* in 2018. "It wasn't until I had an emergency [that I realized]. My office is all the way in the Longworth Building, and I can't run back [before a vote]. Other women showed me a bathroom within Leader Pelosi's office that does have women's products and more privacy, but I shouldn't have to go into her office."

Lori Brown, professor of architecture at Syracuse University and leader of ArchiteXX, a nonprofit organization focused on gender equity in architecture, said, "I was fascinated slash appalled at the lack of access for women in terms of restrooms in the building. It speaks volumes to how much of our public infrastructure and our political infrastructure has been dominated by men ever since its creation."

12. MAKE SURE IT'S ALMOST IMPOSSIBLE FOR HER TO DO HER WORK IF SHE HAS YOUNG CHILDREN

Nursing mothers working on Capitol Hill had no dedicated lactation room until 2006, when the Russell Senate Office Building opened one; in 2007, House Speaker Nancy Pelosi created one in the House buildings. Still, they didn't meet federal standards until 2016. Pregnant staffers were not given temporary parking spaces near their buildings until 2017. In 2018, when changing tables were installed in all the restrooms, female staff and members came to gawp in surprise.

That same year, Senator Tammy Duckworth of Illinois became the first female senator to give birth while in office. Her colleagues voted to change Senate rules so she could bring her infant on the floor during votes. (House rules had allowed lawmakers to bring children on the floor for several years before that.)

The situation for new mothers was far worse in London. The Commons regularly worked late into the night, and while members had a shooting gallery in the basement for a century, where they could blast away at targets to their hearts' content, there was no nursery until 2010. When MP Diane Abbott had her son in 1991, she received no maternity leave and was required to attend Parliament and vote throughout her pregnancy. She was forced to work until four days before she gave birth and was forced to return eight days later. "There was no flexibility, no support, no concern from the whips' office," she said, "you were just expected to turn up and vote." Having no place to leave her newborn, she voted with her son asleep in her arms. The Serjeant at Arms, however, later told her that she had broken the rules and that such an infraction would not be tolerated again. An unnamed Tory MP told *Today*, "This is an outrageous breach of the rules." But MP Don Dixon asked Abbott's critics, "What is Diane supposed to do, leave her baby lying around on the benches?"

"You spent half the time thinking you were a terrible mother," Diane said, "and half the time thinking you were a terrible MP. It was quite stressful, because you didn't feel you were doing either job properly."

13. MAKE SURE SHE KNOWS SHE DOESN'T BELONG THERE

In 1973, when Pat Schroeder arrived on Capitol Hill as a new congresswoman, Speaker of the House Carl Albert congratulated her husband, Jim, on winning the seat and seemed poised to swear him in. "It's her, it's her!" Jim Schroeder said, gesturing with his thumb toward his wife. It took a while for Albert to understand. Albert was not alone in immediately assuming Jim was the new Representative Schroeder from Colorado. Pat Schroeder recalled that her husband "grew weary of saying, 'No, it's her.'" Many male members, baffled, asked Jim, "Why didn't you run?" To which Jim replied, "We ran the strongest candidate." (Jim became a founding member of the

Denis Thatcher Society, a group formed by husbands of power-ful women as a joke and named in honor of Margaret Thatcher's mostly mute husband. Its password was "Yes, dear.")

Lest we think the automatic assumption that the man is the politician has been relegated to the Neolithic era, it happened to Representative Ilhan Omar of Minnesota in 2018 when she showed up with her male chief of staff at the Capitol for orientation. Seated at a small table, the male Capitol Police officer looked only at her associate as he instructed him on safety protocols, completely ig-noring her. When he finished, he shook her associate's hand, wished him luck in Congress, and left.

In 1987, when Nancy Pelosi first entered Congress, there were no female senators and only fourteen women out of 435 represen-tatives. She quickly grew tired of Capitol Police officers stopping her in the corridors and saying, "Sorry, lady, that's for members only." "I am a member," she would say. Again. And again.

One day, as she was following a male colleague, an officer stopped her and told her she couldn't go in there. She had finally had enough. "I can go anywhere I want!" she cried. "I am a member of Congress!"

The officer said, "Congresswoman, that's the men's room."

Pat Schroeder experienced perhaps the most jaw-dropping treatment designed to make it quite clear that she wasn't wanted. When she was assigned to the prestigious Armed Services Com-mittee over the objections of its good-ole-boy chairman, a seventy-two-year-old Louisiana Democrat named F. Edward Hébert, he decided to make sure she knew she didn't belong. Schroeder and a Black congressman, Ron Dellums of California, showed up at their first meeting to find there was only one chair left at the table. Nodding to the chair, Hébert said that women and Blacks were worth only half of one "regular" member, so they'd have to share it. No one else in the room said a word or tried to round up another chair; Hébert could yank funding for a local military base with a snap of his tobacco-stained fingers. Dellums and Schroeder looked at each other and sat down on the chair "cheek to cheek," as Schroeder recalled in her memoir. "Everything in me wanted to rage against this indignity," Dellums later said. "But I thought, let's not give these folks the luxury of seeing that."

"The Lord giveth and the Lord taketh away," Hébert told her, "and here I am the Lord." When she met with him privately to find a way to work together, he said, staring at her, "There are certain people who make me shudder every time they open their mouth." He suggested that she would have greater success on his committee if she used her private parts more (though those weren't the words he used) and her mouth less.

Hébert told Schroeder that women didn't belong on the Armed Services Committee because they knew nothing of combat. She investigated the backgrounds of her male colleagues and discovered most of them had no military experience either, a fact she made known in hearings and the press. Hébert refused to pay for her ticket to an arms control conference in Switzerland. "I wouldn't send you to represent this committee at a dogfight," he told her. She paid her own ticket and went to the press with stories of his horrendous treatment of her and her Black colleague. Hébert lost his prized chairmanship in 1975 and clung on to his seat two more years, a greatly diminished dinosaur.

Unfortunately, the game of musical chairs where the woman loses is not relegated to decades past. On April 7, 2021, Turkish president Recep Tayyip Erdoğan hosted a visit by European Commission president Ursula von der Leyen and European Council president Charles Michel. Von der Leyen oversees laws that affect some 700 million Europeans and, according to protocol, both she and Michel should have been treated equally as heads of state. But when the group walked into the meeting chamber, von der Leyen was shocked to see two armchairs at the head of the room, not three. Erdoğan took the one in front of the Turkish flag, Michel took the one in front of the flag of Europe, and von der Leyen, standing there awkwardly, briefly raised her hands in confusion and said "Ahem." She ended up a tiny figure alone on an enormous sofa, lower than the men, and some twenty feet away from the president, the large gilded chairs, and the flags. The Turkish foreign minister, whose status was lower than von der Leyen's, occupied a similar position on the couch opposite her. Erdoğan had literally put an uppity woman in her place.

It was the "ahem" heard 'round the world; video footage of the incident played on major news stations, and the diplomatic scandal become known as "Sofagate." Twitter exploded with fury against the overt sexist snub with the hashtag #GiveHerASeat. Many users couldn't help but notice that Erdoğan had just pulled out of the Istanbul convention on violence against women. Others tweeted photos of Erdoğan in prior years sitting with the male presidents of the European Council and the European Commission together on identical chairs.

Quite a few criticized Charles Michel for smilingly taking his seat and leaving his colleague stranded rather than asking for another chair and standing, insistent, until one was brought. Or better yet, gesturing for von der Leyen to take the chair next to Erdoğan while smilingly seating himself on the sofa. (Now *that* would have hoisted Erdoğan with his own petard!) But at least Michel insisted von der Leyen be included in the official photo. Erdoğan had planned to keep her out of that, too.

On April 26, von der Leyen spoke about the incident at a meeting of the European Parliament. "I am the first woman to be president of the European commission," she said. "I am the president of the European commission. And this is how I expected to be treated when visiting Turkey two weeks ago, like a commission president—but I was not. I cannot find any justification for [how] I was treated in the European treaties. So I have to conclude that it happened because I am a woman. Would this have happened if I had worn a suit and a tie? In the pictures of previous meetings I did not see any shortage of chairs. But then again, I did not see any women in these pictures, either. . . ."

She continued, "I felt hurt. And I felt alone—as a woman and as a European. Because it is not about seating arrangements or protocol. This goes to the core of who we are. This is what our union stands for. And this shows how far we still have to go before women are treated as equals, always and everywhere. . . ."

Von der Leyen said she was grateful that cameras were in the room when she arrived. "Thanks to them, the short video of my arrival immediately went viral," she pointed out, "and caused headlines around the world. There was no need for subtitles. There was no need for translations; the images spoke for themselves. But we all know, thousands of similar incidents, most of them far more serious, go unobserved, nobody ever sees them, or hears about them, because there is no camera, because there is nobody paying attention. We have to make sure that these stories are told too."

Charles Michel then took the podium and defended himself the best he could, which wasn't all that well, because really what could he say. "I decided not to react further so as not to create a political incident," he said, "that I thought would be still more serious and would risk ruining months of political and diplomatic groundwork made by all our teams at a European level." *In other words, I had a good reason for sitting down on the nice comfy chair and leaving the president of the European Commission standing there like a fool.* He continued, "I would like to reaffirm my total, full and absolute commitment to support women and gender equality."

Except, of course, when a woman really needs a chair.

CHAPTER 13

MISOGYNOIR: WHEN POWERFUL PEOPLE ARE FEMALE AND BLACK

People are not accustomed to a woman, in particular an African American woman, taking this kind of leadership.

—Representative Maxine Waters

Soon after Joe Biden's announcement of Kamala Harris as his running mate, Barry Presgraves, mayor of Luray, Virginia, posted a meme on his Facebook page that said, "Joe Biden just announced Aunt Jemima as his V.P. pick."

For Halloween 2020, Michigan deputy Sherry Prose carved three pumpkins to look like President Donald Trump, Vice President Mike Pence, and President-elect Joe Biden, and depicted Vice President-elect Kamala Harris on a watermelon, a racist trope that emerged in the Jim Crow era.

In the weeks before the announcement, former national security advisor Susan E. Rice—another possible Biden pick—appeared in a meme on a box of Uncle Ben's Rice, labeling it "Uncle Bama's Dirty Rice."

Former first lady Michelle Obama, a Harvard-educated lawyer, was described as an "ape in heels" by Pamela Ramsey Taylor, director of Clay County Development Corp. in West Virginia, who later swore she wasn't racist. In a 2016 interview with a Buffalo newspaper, Carl Paladino, a Trump political ally, said, "I'd like her to return to being a male and let loose in the outback of Zimbabwe where she lives comfortably in a cave with Maxie, the gorilla." Fox News described the first lady as President Obama's "baby mama." Right-wing pundits such as Alex Jones declared that she was a transgender man named Michael Lavaughn Robinson.

Stacey Abrams was also the target of racist robocalls during her campaign for governor of Georgia in 2018. In the call, a voice identifying itself as that of Oprah Winfrey said, "This is the magical negro, Oprah Winfrey, asking you to make my fellow negress, Stacey Abrams, the governor of Georgia." The recording went on to call Abrams a "poor-man's Aunt Jemima."

This vicious diminishing of Black women is called "misogynoir" (*noir* meaning "black" in French,) a term coined by Moya Bailey, an African American feminist scholar, for what happens at the intersection of sexism and racism. And, just as calling out misogyny is usually seen as a more egregious crime than the misogyny itself, pointing out racism is usually considered far more appalling than the racism itself. In his 1992 study "Discourse and the Denial of Racism," University of Amsterdam professor Teun van Dijk found that, "Accusations of racism . . . tend to be seen as more serious social infractions than racist attitudes or actions themselves." (As in, a substantial portion of the American population think George Floyd: meh; Critical Race Theory: outrage!)

Black women suffer all the tried-and-true measures in the Misogynist's Handbook to keep women down and much more. Let's take ambition, for example. Black women are often harshly criticized for making known their ambitions. Let us recall how Kamala Harris was criticized for being overly ambitious in aiming for the presidency. And how Stacey Abrams was called "offensive," "inappropriate," "entitled," "desperate," and "obsessively ambitious" for aiming for the vice presidency.

And yet, many Black women will be overlooked if they do not make their ambitions quite clear to those who hold the key to promotion—usually white men—according to LaTosha Brown, cofounder of Black Voters Matter, an organization focused on increasing Black voter turnout. "Ambitious women have always been a problem for those who have wanted to maintain the status quo—the White male patriarchal power structure," she told the *Fix*. "The interesting piece though about this is that if Black women didn't have a measure of ambition, there is no way that we would be able to navigate the highly patriarchal environment that we've been forced to endure since arriving on these shores as enslaved Africans."

Let us examine the subject of emotions. We've seen that the Misogynist's Handbook paints women as emotional wrecks pulsating with dangerous hormones: hysterical, weeping, prone to erratic outbursts. But there is a particular adjective reserved for Black women: "angry." Soon after Joe Biden announced Kamala Harris as his VP pick, Donald Trump referred to her numerous times as "angry," "extraordinarily nasty," and "a mad woman," racist-coded descriptions. Recalling her pointed cross-examination of Brett Kavanaugh during his confirmation hearings, Trump said that she was "so angry and [had] such hatred with Justice Kavanaugh . . . she was the angriest of the group." "She left [the presidential race] angry," Trump said. "She left mad."

Georgetown University professor and author Michael Eric Dyson told the *New York Times* in October 2020, "The notion of the angry Black woman was a way—is a way—of trying to keep in place Black women who have stepped outside of their bounds, and who have refused to concede the legitimacy of being a docile being in the face of white power."

While anger at injustice has fueled great social and political movements—the American Revolution, for instance, the Civil Rights Movement, and #MeToo—many Black women accused of anger are not angry in that moment at all; they are simply speaking. And when Black women point out the injustice of being labeled "angry," they just appear angrier to those who call them such. The easiest choice is to remain silent, which, though understandable on the personal level, just lets misogynoirists win, as silencing Black women is the very purpose of the creation of the angry Black woman trope.

"HANG THESE TRAITORS WHERE THEY STAND"

While many female politicians must put up with *Lock her up!*, women of color must also endure *Send her back!* The message of not belonging, of needing to go back, has been directed at four young, left-leaning women of color elected to Congress in 2018 known as "the Squad." Alexandria Ocasio-Cortez of New York is of Puerto Rican descent. Ayanna Pressley is the first Black congresswoman to represent Massachusetts. And Ilhan Omar of Minnesota and Rashida Tlaib of Michigan are the first two Muslim women ever elected to Congress. Tlaib is Palestinian American, and Omar was born in Somalia. That the four are female and Brown and two are Muslim fuels the perfect storm of abuse.

It's easy to understand why the Squad's progressive politics would anger Donald Trump and his supporters, but instead of focusing on their politics, he falsely implied that the women weren't American; all are American citizens, three of them were born here, and Omar came as a child. In July 2019, Trump tweeted they should "go back and help fix the totally broken and crime infested places from which they came." Trump supporters at a rally in North Carolina responded to his criticisms of Omar by chanting, "Send her back!"

In 2019, George Lakoff, a cognitive linguist, told the *Washington Post*, "'Send her back' has the same grammatical structure as 'Lock her up,' and the same sound structure—it's very straightforward, and it has virtually the same meaning."

These women have also been on the receiving end of various versions of "Off with her head!" In 2020, Twitter suspended the account of one of Omar's Republican challengers, Danielle Stella, for calling for Omar to be tried for treason and hanged if she had, as a conspiracy theory stated, given top-secret material to Iran. Stella later tweeted a link to a stick figure hanging from a gallows. George

Buck, a Republican challenger to Democratic representative Charlie Crist in Florida, accused Omar of giving information to Qatar and stated, "We should hang these traitors where they stand."

Amanda Hunter, executive director of the Barbara Lee Family Foundation, believes that hatred of the Squad is rooted in the fear of their power. "When you look at the Squad in Congress, the women who receive the most vitriol, they are the same women who inspire and motivate the most people in the country," she said in an interview for this book. "Their time in Congress has been relatively short, and yet their national standing is so high. Any time I ask younger women whom their favorite elected officials are, they say AOC is one of their heroes. It is important not to underestimate the power these women have, which can be scary to white men."

THE UNBOUGHT AND UNBOSSED JOURNEY OF SHIRLEY CHISHOLM

Shirley Chisholm, the first Black woman elected to Congress in 1968, and the first Black candidate for a major party's nomination in 1972, didn't shrink from taking on the double bias that came her way as a Black woman. "If they don't give you a seat at the table, bring a folding chair," she advised. Chisholm perceived many similarities between racism and sexism. "The cheerful old darky on the plantation and the happy little homemaker are equally stereotypes drawn by prejudice," she wrote in her memoir.

Of the difficulties she faced as a Black woman, she found sexism to be worse than racism and often encountered it from Black males as well as white ones. "I met far more discrimination being a woman than being black when I moved out into the political arena," Chisholm noted. "Of my two 'handicaps,' being female put many more obstacles in my path than being black. Sometimes I have trouble, myself, believing that I made it this far against the odds."

Having received a master's degree in elementary education from Columbia University in 1952, Chisholm worked as a daycare director for many years. She became involved in local politics in 1953, serving as a volunteer to promote civil rights and economic opportunities in Brooklyn. After ten years of helping men win public office, in 1964 Chisholm decided to run for a New York State Assembly seat and won. In 1968, when a new congressional district in New York was created, she decided to throw her hat in the ring. She thought voters would like the fact that she was independent and didn't owe any of the powerful local party bosses anything. Her campaign slogan was, "Shirley Chisholm: Unbought and Unbossed."

Chisholm won the primary. In the general election, she ran against James Farmer, a colleague of the Rev. Dr. Martin Luther King Jr. and a cofounder of the Congress of Racial Equality. He'd helped organize lunch counter sit-in protests and Freedom Rides that challenged segregation in interstate travel. But Farmer, a Black liberal candidate running as a Republican, attacked Chisholm for being a woman. "Women have been in the driver's seat in Black communities for too long," Farmer said. He argued the district needed "a man's voice in Washington," not that of a "little schoolteacher."

Chisholm pushed back. "There were Negro men in office here

before I came in five years ago, but they didn't deliver," she countered. "People came and asked me to do something. . . . I'm here because of the vacuum."

During the campaign, Chisholm was diagnosed with a massive tumor in her abdomen and required emergency surgery. The doctor wanted her to rest for weeks afterward, but her opponent was ridiculing her absence as female weakness in the press. "Look," she told the doctor, "the stitches aren't in my mouth. I'm going out."

She recalled, "I took a big beach towel and wrapped it around my hips so my clothes wouldn't fall off. With that, I looked pretty good. I bribed two women to help and three men. We lived on the third floor then, and I had to walk down three flights. I told the biggest one, 'You walk in front so if I fall I'll fall on you and the other two can hold me.'"

On the back of a truck she spoke through a megaphone, "Ladies and gentlemen, this is Fighting Shirley Chisholm and I'm up and around in spite of what people are saying." Chisholm beat Farmer in 1968 by getting out the women's vote. There were only nine Black members of Congress when she joined, all of them men.

New members are often assigned to uninteresting committees and work their way up over time. Even so, Chisholm was shocked that a congresswoman from Brooklyn would be assigned to the Agriculture Committee. "Apparently all they know in Washington about Brooklyn was a tree grew there," she later said, adding that the only crop grown in Brooklyn was marijuana. Chisholm met with Speaker John McCormack to ask him to change her assignment to one with greater relevance to her district. He refused. She told him she would do what she needed to do.

At the next session, Chisholm kept standing up, expecting to be called on. After six or seven attempts, she walked down to the Speaker's dais and was recognized. "I would just like to tell the caucus why I vehemently reject my committee assignment," she said. "I think it would be hard to imagine an assignment that is less relevant to my background or to the needs of the predominantly black and Puerto Rican people who elected me, many of whom are unemployed, hungry and badly housed, than the one I was given." She asked for a new assignment and was later given veterans' affairs. "There are a lot more veterans in my district than there are trees," she said. The New York *Daily News* praised her courage.

In Congress, Chisholm advocated for guaranteed minimum annual income for families. She pushed for extended hours at daycare facilities. She supported national school lunches. She resented that the Vietnam War took much-needed money from housing and food programs and Head Start, which helped poor children get a jump on education. In her first speech from the House floor, on March 26, 1969, Chisholm criticized the war, calling the US hypocritical for its international diplomacy of trying to "make the world free" when racism raged at home.

In 1972, she made a revolutionary decision: she would run for president. She knew she wouldn't win. But she also knew there had to be a Black woman leading the way for others to follow. "I sought the presidency so the next time a woman or a black person decides to make a bid for the presidency," she wrote in her memoir, "that that individual will not have to be on the defense for five months just because he is black or because she is woman; that this is a multifaceted society that should be able to mobilize the talents of all kinds of citizens, and traditionally, because the presidency has been the exclusive domain of white males, people laughed at the idea of anyone other than a white male running for the presidency of the United States as a fool. I blazed the trail. I went to the edge so that now any black or any woman running will not be regarded as some folly or some evil."

Chisholm wrote that the women she knew in government seemed to have a stronger moral purpose and were less inclined to wheel and deal. "A larger proportion of women in Congress and every other legislative body would serve as a reminder that the real purpose of politicians is to work for the people."

In response to accusations that she was biased against men and whites, she wrote, "I am not anti-male any more than I am anti-white, and I am not anti-white, because I understand that white people, like black ones, are victims of a racist society. They are products of their time and place. It's the same with men. This society is as anti-woman as it is anti-black. It has forced males to adopt discriminatory attitudes toward females. Getting rid of them will be very hard for most men—too hard, for many of them."

In 1982, Chisholm announced she would not seek reelection to Congress. "I'm hanging up my hat," she announced. She moved to Palm Coast, Florida, where she continued to lecture and write. In 1993, President Bill Clinton nominated Chisholm to become US ambassador to Jamaica, but she withdrew because of ill health.

Chisholm, who died January 1, 2005, wrote, "I hope if I am remembered it will finally be for what I have done, not for what I happen to be. And I hope that my having made it, the hard way, can be some kind of inspiration, particularly to women."

"YOU MUST HAVE COME TO DO THE WASHING UP"

By the time thirty-four-year-old Diane Abbott became the first Black female MP in British history in 1987, she had already had her fair share of racist misogyny. For instance, there was the occasion when she attended a glamorous ball as an undergraduate at the University of Cambridge. "I was dressed up in a long evening dress and made up and bejeweled to within an inch of my life," she wrote in an article in the *Times* in 1997. "Yet as soon as I came in through the gate someone rushed up to me and said, 'Oh good, you must have come to do the washing up.' He did not ask himself why I would wear an evening dress and diamante to do so. He only knew that I was a black woman and therefore must belong in the kitchen." When she served on the Westminster City Council in 1982, the security guards at the Council House tried to turn her away.

Abbott arrived in Parliament with three Black male friends who had won seats in the same election. "One of the things we found when we first entered Parliament was that none of the attendants believed we were MPs," she told her biographers in 2017. The Serjeant at Arms and the security staff frequently asked them what they were doing there or blocked them from going where they wanted to go.

Security staff were also unwilling to let their Black visitors into the building. Abbott held events in Parliament to support Black groups, but they often got off to a late start because her guests were prevented from entering. In April 1988, she and her three Black male colleagues sent a letter to the parliamentary authorities making the problem known. "Ever since my colleagues and I have been in the House there have been a series of incidents that give rise to concern," they wrote. "Our visitors are sometimes treated less than politely and deliberately misled. . . . Visitors and we ourselves have been jostled. We have been challenged by attendants as to our identity in an unsubtle attempt to embarrass us." When the letter didn't seem to get them anywhere, she went public with the accusations, speaking to the media.

In the early 1990s, Abbott's brother, a civil engineer, attended an Institution of Civil Engineers dinner at the House of Lords. Chatting over dinner, he mentioned that his sister worked in Parliament. "So she works in the kitchen?" came the reply.

"As a black woman MP, you can face two things," Abbott said in a 2011 interview. "You face sexism—men not wanting to take you seriously, and people generally taking men more seriously than you. You also face racism—people feel you can't be as good, you can't be as competent."

As a result of misogynoir, Abbott has been inundated with abusive tweets. In 2017, Conservative councilor Alan Pearmain tweeted an image of an orangutan photoshopped to look as though it was wearing lipstick, with the caption, "Forget the London look, get the Diane Abbott look." Pearmain defended the tweet, noting, "People will take offence about everything, won't they?"

In 2016, the *Telegraph* wrote about an extract on a biography of opposition leader Jeremy Corbyn soon to be published. Back in 1987, the paper reported, when Abbott lived briefly with Corbyn, he had driven two friends to the apartment they shared to show them Diane naked in his bed. It turned out the book reported no such a thing. Soon articles suggested Abbott had only been given a job in the shadow Cabinet because of her relationship with Corbyn forty years earlier. That after serving thirty years in Parliament, she had not earned the position but had been rewarded with it for sexual favors in the distant past.

In 1985, Abbott wrote in the *West Indian World*, "I find white people will actually tolerate and even encourage any black person who they think they can control or who they do not regard as intelligent, but if they think you have a mind of your own, they feel very threatened. . . . As a black woman you come under particular pressure. Most white people find it very difficult to accept a black woman in a position of authority."

"SHE SEEMS LIKE A GREAT HOUSEKEEPER"

Italy's first Black government minister, Cécile Kyenge, was born in the Congo. She immigrated to Italy at the age of nineteen in 1983, where she studied medicine and became an ophthalmologist. From 2013 to 2014, she served as minister for integration, helping to assimilate immigrants who now make up about 7 percent of the population, around four million people. From 2014 to 2019, she served as an Italian member of the European Parliament.

Kyenge has been subjected to vicious racist attacks. In 2013, the far-right party Forza Nuova dumped three mannequins stained with fake blood outside a town hall where she was due to make a speech. Forza Nuova member Pablo De Luca accused Kyenge of planning "the destruction of the national identity." He said, "Her words overflow with racism against European culture." A well-known Italian winemaker, Fulvio Bressan, called her a "dirty black monkey."

A former vice president of the Italian Senate, Roberto Calderoli, said in a public meeting, "When I see pictures of Kyenge I can't help but think of the features of an orangutan." When questioned about the comments, Kyenge said she would not demand Calderoli's resignation, but she encouraged politicians to "reflect on their use of communication." She told an Italian news agency, "I do not take Calderoli's words as a personal insult, but they sadden me because of the image they give of Italy."

Mario Borghezio, a member of the European Parliament, said he feared Kyenge would impose "tribal conditions" on Italy and help form a "bongo-bongo" administration. "She seems like a great housekeeper," he added. "But not a government minister." He helpfully pointed out that Africa had "not produced great genes."

"Other extreme-right politicians have called me 'Zulu' and 'Congolese monkey,'" she wrote in the *Guardian* in 2018. "I have faced death threats and now live under police protection."

In 2013, someone in the audience threw bananas at her while she spoke. They fell just short of the podium, and she ignored them. But later, she tweeted, "With so many people dying of hunger, wasting food like this is so sad."

"I'M NOT ANGRY AS MUCH AS I AM DETERMINED"

One Black female politician who is often labeled as angry is Maxine Waters. Born in 1938 "too skinny," "too black," and the spitting image of the father who abandoned her family, she was the fifth of thirteen children raised by a mother struggling financially. "Just getting *heard* in a family that size is difficult," she told *Ebony*, which probably explains her talent for oratory. In 1976, she was elected to the California State Assembly, where she successfully pushed for the state to divest from South Africa's apartheid regime. She was elected to Congress in 1991, where she vociferously opposed the Iraq War.

Dubbed "Kerosene Maxine" and "Mad Max" by opponents using the angry Black woman trope, in 1994, Waters got into a shouting match with Republican congressman Peter King over whether he was badgering a female witness during a hearing on the Whitewater controversy. She felt that men giving testimony were treated respectfully, whereas King had treated the woman rudely. King told her to sit down. She told him to shut up. The following day, he raised the issue on the floor, angrily decrying her behavior.

Waters strode to the podium and said, "Thank you very much, Madam Chairwoman. Last evening a member of this house, Peter King, had to be gaveled out of order at the Whitewater hearings of the Banking Committee. He had to be gaveled out of order because he badgered a woman who was a witness from the White House, Maggie Williams. I'm pleased I was able to come to her defense. Madam Chairwoman, the day is over when men can badger and intimidate women, marginalize them, and keep them from speaking."

Calls were immediately made from the floor to stop her from speaking and to strike her words from the record. Representative Carrie Meek of Florida, who was presiding over the chamber, kept banging her gavel, crying, "You must suspend!" Waters kept going, not shouting, but speaking loud and clear. "I am pleased I was able to come to her defense. We are now in this House," she said. "We are members of this House. We will not allow men to intimidate us and to keep us from participating."

The floor descended into chaos. The chair kept banging her gavel. The men were losing their minds that a Black woman was calling them sexist. They shouted to punish her, to adjourn the House. Someone called for the mace, a giant forty-two-inch-tall magic wand of black rods made in 1841, topped by a winged eagle on a golden globe. Merely holding the mace before a troublemaker on the House floor is supposed to have the effect of crying "Silencio!" in a Harry Potter novel. (To which she would have replied, "Expelliarmus!") There was some confusion about what the mace was, and where it was (it was right behind the podium, leaning against the wall), and what to do with it when they found it because it hadn't been used since 1917. But thirty-five seconds of a Black woman decrying sexism was enough for frantic calls to find it fast, dust it off, and wave it violently in front of her mouth.

"Do you ever see men do this to other men?" Waters continued as the gavel pounded and the men shouted. "This is a fine example of what they try to do to us. The women of this nation will not continue to have this kind of treatment. Thank you, Madam Chairwoman." She abruptly left the podium.

"Have the Sergeant-at-arms remove her!" cried one congressman. Another called for her to be "maced" even though she had left the floor. Waters had clearly struck a nerve. Tom Foley, Speaker of the House, took over to sort out the mess. He said, "While in the opinion of the chair, while the words were not in themselves unparliamentary, the chair believes that the demeanor of the gentle-woman from California was not in good order," and suspended her from the floor for the rest of the day.

So it wasn't what she said that was wrong. It was her . . . bearing?

Pat Schroeder of Colorado, who by that time had experienced two decades of sexism in Congress, jumped to Waters's defense. "Mr. Speaker," she said, "I'm a little puzzled at the word 'de-meanor.'" She believed Waters couldn't hear the chair asking her to suspend what with all the men yelling and shouting. "How can you challenge 'demeanor'?" Schroeder asked. Foley replied that Waters should have stopped talking when asked. Her odious words were struck from the record. Later, Waters said she thought the chair was telling the shouting men to suspend, not her, and she never intended to disobey.

"Women are new to this place," she told the *Los Angeles Times* soon after the fracas. "Women are supposed to know their place. I exercise my rights, and it's new for men. It's not easy for them to accept women as equal partners." In 2017, after a contretemps with President Donald Trump, she told CNN, "People are not ac-customed to a woman, in particular an African American woman, taking this kind of leadership."

Is Maxine Waters angry? In 2018, she told *Elle* magazine, "I am an experienced legislator, who understands strategy, who under-stands the value of speaking truth to power, and I'm not angry as much as I am determined."

In 2017, she had five minutes to question the new secretary of the treasury, Steven Mnuchin, during a meeting of the House Fi-nancial Services Committee. When she asked him why he did not respond to a letter she had sent him two months earlier regarding President Trump's financial ties to Russia, he started thanking her for her service, spinning out the time so he wouldn't have to answer her. "Reclaiming my time!" she cried, over and over again as Mnuchin dithered and dawdled and seemed confused. The video went viral and inspired a gospel-style song.

In 2017, after Waters denounced Donald Trump on the House floor, Bill O'Reilly of Fox News was asked what he thought about her speech. Evidently, he couldn't think of anything intelligent to say, so he replied, "I didn't hear a word she said. I was looking at the James Brown wig." He was indicating she looked like a man, another tired, old insult to Black women.

In response, Waters told MSNBC's Chris Hayes, "Let me just say I'm a strong black woman and I cannot be intimidated. I cannot be undermined. I cannot be thought to be afraid of Bill O'Reilly or anybody. And I'd like to say to women out there everywhere: Don't allow these right-wing talking heads, these dishonorable people, to intimidate you or scare you. Be who you are. Do what you do. And let us get on with discussing the real issues of this country."

"COULD YOU REPEAT THAT QUESTION?"

Twice before the 2020 election, a presidential nominee chose a woman as his running mate in a desperate move to gin up flagging support. In 1984, Democratic nominee Walter Mondale picked New York congresswoman Geraldine Ferraro—the first female vice presidential candidate ever. Incumbent president Ronald Reagan was so extremely popular, Mondale needed to think outside the box to win voters. Ferraro—well-spoken, attractive, and competent—could win over millions of female voters. It wasn't enough. Mondale and Ferraro lost in a landslide.

It took until 2008 for a woman to appear on a major-party ticket again, when Republican nominee John McCain named Sarah Palin as his running mate—and once more it was a desperation move, as McCain faced the inspirational Democratic nominee Barack Obama. Attractive and as appealing as a breath of fresh air—at least initially—she ended up harming, rather than helping, the campaign.

But in August 2020, when Joe Biden selected Kamala Harris as his running mate, it was a strategic choice. "Harris got picked when it looked like Biden had a chance to win—it wasn't just a desperation move," Joanna Howes, head of the Women's Vote Project, told the *Washington Post* November 1 of that year. "What I think was impressive this time is how many women there were as opposed to 1984. This time, there were governors, there were senators, there were members of Congress."

Shaunna Thomas, cofounder of the feminist organization UltraViolet, noted the reaction of the Trump campaign to Harris's selection. "There are immediately, out of the gate, sexist and misogynistic and racist attacks having absolutely nothing to do with her record or the substance of her leadership," she told the *Washington Post* on August 12. It was less effective for Trump, an older white man, to attack Biden, another older white man. True, Trump called him "sleepy," accused him of hiding in his basement during the pandemic, and suggested he had dementia. But those criticisms were anemic. The vicious attacks were reserved for the Black woman on the ticket, using the Misogynist's Handbook as a guide.

A Black female vice presidential candidate can expect "disrespect that is a dual assault on their race and gender," Errin Haines, an editor at the news site the *19th*, told CNN the day before Biden announced his selection of Harris. "She can expect to be attacked, vilified, and criticized for daring to have ambition, capability and a voice in American politics."

In addition to the racist trope of being angry, and the misogynistic tropes of being alarmingly ambitious and vaguely unlikable, Harris also had to put up with the birtherism that had been lobbed a few years earlier at President Obama. Though she was indisputably born in Oakland, California, opponents said she was not really a US citizen because her parents weren't citizens at the time of her birth, totally ignoring the Fourteenth Amendment to the Constitution. In other words, Harris didn't belong here. Another attack came from conservatives such as Pat Robertson, Rush Limbaugh, and Dinesh D'Souza who claimed that she was not Black enough to call herself Black: her mother was from India and her father is a Jamaican of African descent, and her calling herself Black was a clear sign of inauthenticity. But even if she wasn't Black enough, she was an angry, nasty Black woman.

It was easier to lob sexist and racist tropes at Harris than attack her impressive background. Armed with a law degree, in 1990, Harris joined the Alameda County District Attorney's Office prosecuting child sexual assault cases before serving as a managing attorney and chief of the Division on Children and Families in the San Francisco District Attorney's Office. In 2003, she was elected San Francisco District Attorney. Eager to assist first-time drug offenders to stay off the street, she formed a trailblazing program to offer them the opportunity to earn a high school degree and find a job. The US Department of Justice named Harris's program a national model of innovation for law enforcement.

In 2010, Harris was elected California's Attorney General, directing the largest state justice department in the country. Refusing to accept a portion of the measly settlement the big banks were offering Californians who had lost their homes in the 2008–2009 economic crisis, she won a $20 billion settlement. She was elected California senator in 2018, joining the Senate Select Committee on Intelligence, where she worked to protect the US against foreign threats. Her sharp, relentless interrogation of witnesses during several nationally televised hearings got her noticed by other politicians and the American public at large. Perhaps her most impressive moment was questioning Attorney General Bill Barr with regard to the Mueller investigation in May 2019.

"Attorney General Barr, has the president or anyone at the White House ever asked or suggested that you open an investigation of anyone?" she asked.

"Um, I wouldn't . . . I wouldn't, uh . . ." Barr babbled.

"Yes, or no?" She narrowed her eyes.

"Could you repeat that question?" he asked, looking helplessly around.

"I will repeat it. Has the president or anyone at the White House ever asked or suggested that you open an investigation of anyone? Yes or no, please, sir."

"Umm. The president or anybody else . . ."

"Seems you'd remember something like that and be able to tell us?"

"Yeah, but I'm trying to grapple with the word 'suggest.' I mean, there have been discussions of matters out there," he waved his beefy hand around airily, "that. . . . They have not *asked* me to open an investigation."

"Perhaps they suggested?" She nodded and narrowed her eyes again.

"I don't know. I wouldn't say 'suggest.'"

"Hinted?"

"I don't know."

"Inferred?"

At this point, Barr gave up answering altogether and made a funny mouth.

"You don't know," Harris said. "Okay."

CHAPTER 14

OFF WITH HER HEAD!

Men are afraid that women will laugh at them.
Women are afraid that men will kill them.

—Margaret Atwood

When the Misogynist's Handbook marches relentlessly forward unimpeded over an inconvenient woman who refuses to sink back into her place, the result can be death. One of the most powerful images of female death in ancient mythology is that of the hero Perseus holding up the head of snake-haired, monstrous Medusa. Virtue has triumphed over evil. A man has firmly put a troublemaking woman in her place; he has vanquished her female power and returned the world to the comforting safety of men wielding swords.

There is far more to the Medusa story than meets the eye. Medusa was originally a North African goddess of women's wisdom. Snakes—living deep in the womb of the earth—were revered for their connection to sacred female power. They were also symbolic of healing and renewal. Shedding their skins, they emerge seemingly younger and healthier. (There is a reason that in 1910 the American Medical Association chose as its symbol the rod of Asclepius, the ancient Greek god of healing, with two snakes spiraling up, the choice a prescient mirroring of double-helix DNA.)

For many centuries, divine snake power played a major role in many Mediterranean religions. The womblike Maltese temples dedicated to the sacred fat lady are covered in snake spirals. Minoan snake goddess statuettes from around 1600 BCE hold up a serpent in each hand, possibly as a symbol of feminine power. The Egyptian goddess of sacred ecstasy and sexual pleasure, Qetesh, wears a moon on her head and holds a snake. The temples of Astarte, Phoenician goddess of sexuality and fertility, were decorated with snakes. The Furies, female spirits of vengeance, were often depicted as having snakes for hair. The witch Medea flew in a chariot pulled by serpents. The man-killing Greek maenads wore living snakes as jewelry. The basilisk, a snake whose very glance killed, was born of menstrual blood.

The famous Oracle of Delphi in Greece, dedicated to the god Apollo, had originally been a shrine to a giant snake, Python, who was said to utter prophecies, starting in about 1400 BCE. The name Delphi—which means *womb*—may signify that the primordial goddess Gaia, the ancestral mother of all life, was worshipped there. But some seven centuries later, Apollo (or, more likely, his priests) killed Python, took over the joint for himself, and every four years celebrated the massacre by holding testosterone-filled sporting events—the Pythian Games, where men beat the crap out of each other and raced chariots around in circles. Apollo named his chief priestess *Pythia* after the snake, but she was firmly under male control.

According to mythologist Joseph Campbell, the story of Medusa's murder—and, it is fair to say, that of Python—was created by invading men to justify their aggression. "Wherever the Greeks came," he wrote, "in every valley, every isle, and every cove, there was a local manifestation of the goddess-mother." To obtain power for themselves, they needed to cut off her head. Apollo slew Python. Perseus beheaded Medusa. The baby Hercules strangled two snakes the jealous goddess Hera sent to kill her husband's illegitimate son. The writers of the biblical Book of Genesis ensured that snakes were seen as deceitful, manipulative, and evil for all time, which is why Saint Patrick drove them out of Ireland to eternal acclaim. These tales tell of the emphatic rejection of divine feminine power, replaced by a man dressed in battle armor swinging a mace.

One aspect of the Medusa story most people don't know is that the god Neptune raped her in the goddess Minerva's temple, and Minerva, angry at the defilement, blamed the victim—some things never change—not the rapist, and changed her into a snake-haired monster whose glance could turn people to stone. Curiously, snakes represent not only the feminine divine but, given their phallic shape, also penises. It's not a giant leap to imagine Medusa with writhing penises for hair. Clearly, she—and all those other snaky women—represented not only scary female power, but also castration, which to some men is probably the same thing. (Where, good God, did all those writhing penises on Medusa's head come from? And did she feed them oats and corn?) Perseus, by cutting off the head of the castrator—the woman weakening men either physically through sexual desire or metaphorically by taking their power—has saved the Patriarchy. Tucker Carlson must be glad.

The compelling symbolism of Medusa has never faded. Marie Antoinette was frequently portrayed as a monster with snakes for hair. In the nineteenth century, women's rights activist Susan B. Anthony recognized that the world was still chock-full of Medusas and Perseuses. "Women must echo the sentiment of these men," she wrote. "And if they do not do that, their heads are cut off."

Many women running for high political offices have been portrayed online as Medusa cut off at the neck, including Angela Merkel, Theresa May (labeled "Maydusa" in the meme), and Elizabeth Warren. But perhaps most disturbing is a 2016 meme of Donald Trump's face grafted onto the youthful, muscular body of Benvenuto Cellini's 1554 statue of Perseus, holding up Hillary Clinton's hideously grinning head by her snaky hair. Soon after Biden's announcement of Harris as his running mate, an image of her with writhing serpent hair made it onto social media. Reaction to such images is usually muted. It is such an old trope it tends to make one yawn.

But in May 2017, when comedian Kathy Griffin posted a photo of herself holding up a Donald Trump mask made to look like a severed head, she was fired from CNN, blacklisted, threatened with being charged with conspiracy to assassinate the president, and banned from flying for two months because her name was on the no-fly list along with all the terrorists. Off with *his* head is simply not allowed.

If a literal beheading is going too far in our supposedly civilized society, there's always the option of locking her up. In 1872,

Victoria Woodhull was the first women ever to run for US president. In *The Highest Glass Ceiling: Women's Quest for the American Presidency*, author Ellen Fitzpatrick wrote, "Ambition alone was alienating to some and her most vociferous critics . . . even likened her to the devil. Rather than send her to the White House, there [were] those that wished to see her locked up in prison on election day."

Nearly a century and a half later, in 2016, the same sentiment held true for Hillary Clinton at the Republican National Convention, where she was likened to the devil and threatened with jail time. "We know she enjoys her pantsuits. . . . What she deserves is a bright orange jumpsuit!" shouted Darryl Glenn, Colorado's Republican Senate candidate, to riotous applause. On the second day of the convention, New Jersey governor Chris Christie played prosecutor at a mock trial, calling out each of Clinton's "crimes" to delighted shouts of "Guilty! Guilty!" from the audience. The crowds took every opportunity to chant "Lock her up!" At a rally in June 2016, Trump supporters cried, "Hang her!"

While Clinton managed to cling onto her head, throughout history, some powerful women literally did lose theirs: Anne Boleyn; Mary, Queen of Scots; and Marie Antoinette, while Hypatia was torn to little pieces and set on fire, and Cleopatra, though remaining in one piece, lost her life. But lest we believe that murdering powerful women is strictly a thing of the long-ago past, let's look at recent history.

In 1984, police arrested a maintenance man at a company where vice presidential candidate Geraldine Ferraro would be speaking for planning to shoot her with a bow and arrow. They found the weapons—along with a pistol—in the trunk of his car. The would-be assassin felt that a woman should not be vice president.

While meeting with constituents in 2011, forty-year-old Arizona congresswoman Gabrielle Giffords was shot in the head in a mass shooting event that killed six others, including a nine-year-old girl. Though Giffords survived, she still has difficulty speaking and walking and has lost 50 percent of her vision in both eyes. Her assailant believed that women should not hold positions of power. In 2016, thirty-eight-year-old British MP Jo Cox was shot and stabbed to death in the street while on her way to meet constituents by a man who held extreme far-right views.

On March 14, 2018, thirty-eight-year-old Marielle Franco, a city councilor of the Municipal Chamber of Rio de Janeiro for the Socialism and Liberty Party, was assassinated, along with her driver in her car. Franco was Black, gay, from a poor favela, and fought against police violence, LGBTQ violence, and gender violence, and campaigned for reproductive rights and the rights of favela residents. One suspect, a police officer, was killed while attempting to resist arrest, the story goes. Two other former officers sit in jail awaiting trial, and five people have been charged with obstructing justice by hiding evidence. Evidence seems to point at the hit being ordered by someone at a high level in government. While Franco was especially despised for being an outspoken gay female, it is difficult to say whether those attributes played a role in her murder. Within five months, ten other Brazilian political activists were murdered, and they were all men.

Indeed, one means of determining whether abuse is gender-based is to compare the threats to female and male politicians and activists in a particular country, consider the general culture of violence, and look into the motives of the attackers. For instance, the assassinations of Benazir Bhutto in Pakistan and Indira Gandhi in India seem to be unrelated to their gender.

We must also examine whether there is a huge difference in the kinds of threats aimed at men and women. For instance, are women threatened with being skinned alive, dismembered, dipped in acid, torn into little pieces, and called gender-coded insults like *cow*, *pig*, *dog*, *bitch*, and *cunt*? For instance, Katharina Schulze, co-leader of the Greens party in Bavaria, said that some 20 percent of her emails were abusive, many threatening her with rape, according to a 2019 BBC *Newsnight* investigation. Her male Greens co-leader Ludwig Hartmann, who espouses the exact same policies, receives messages calling him a communist.

First Lady of Namibia Monica Geingos also noticed the gender-based differences in the insults flung at her and her husband, President Hage Geingob. Online trollers called him an "oxymoron nincompoop." In a video she released on International Women's Day, March 8, 2021, Geingos said, "I don't know what the hell an oxymoron nincompoop is," (and I imagine most of us don't either) "but why can't I also be a neutral insult like an oxymoron nincompoop? I also want to be a nincompoop. I don't want to be a gold digger, a slut, a bad mother, a Jezebel. I don't want to be asked when I am having a baby, to be told I am too ambitious, too loud, that I should shut up."

Every weekday morning, one of the first tasks of the staff of Diane Abbott, the first Black woman elected to the British Parliament, is to delete and block abusive messages, "usually while having breakfast," said one staffer in a 2017 report to Parliament called "Intimidation in Public Life." "Porridge in one hand, deleting abuse with the other." Those that seem truly disturbing are turned over to the police.

During a 158-day study in 2017, Amnesty International found that of the 650 members of Parliament, Abbott was the target of almost a third of abusive tweets, a figure that rose to more than 45 percent in the weeks leading up to a general election. That's an average of fifty-one threatening tweets per day.

Abbott told the Amnesty researchers, "It's the volume of it which makes it so debilitating, so corrosive, and so upsetting. It's the sheer volume. And the sheer level of hatred that people are showing. . . . It's highly racialized and it's also gendered because people talk about rape and they talk about my physical appearance in a way they wouldn't talk about a man. I'm abused as a female politician and I'm abused as a black politician."

"I've had death threats," she told Parliament on July 12, 2017. "I've had people tweeting that I should be hung if 'they could find a tree big enough to take the fat bitch's weight'. . . . I've had rape threats, been described as a pathetic, useless, fat, black piece of shit and an ugly, fat black bitch, and n——, n—— over and over again."

When Abbott spoke with a policeman working on Jo Cox's murder, she learned that the assailant had papered a room with photos of his victim. Abbott thought, "I have no doubt that there's someone out there with a whole wall papered with pictures of me." She found it particularly disturbing when police, who had not acted on her death threats, arrested a man for threatening to kill a white female MP.

Social media is a weapon of the Patriarchy, a potent new tool in the Misogynist's Handbook to threaten, abuse, and belittle powerful women. It took decades for the first whispers to come out about Isabeau of Bavaria, centuries for them to cement themselves in the public consciousness. No longer do *libellistes* manually print scandalous pamphlets in a foreign country, smuggle them into a capital city in a nobleman's baggage, and sell them surreptitiously in darkened bookstores, as they did to take down Marie Antoinette. Nor do misogynists sit down, write a letter, stuff it into an envelope, address it, put a stamp on it, and walk it to the mailbox. These days, with just the push of a button . . . *whoosh!* It's gone around the world, viral in hours, viciously punishing a woman for stepping outside of patriarchal bounds.

Online harassment is a form of public gender-role enforcement, rather like dragging a loud-mouthed woman to the stocks in the town square, clamping a scold's bridle on her mouth, and throwing rotten vegetables at her. Except these days abusers can do it sitting at home, drinking coffee, and completely anonymously. Its purpose is to silence her, to force her to conform. To let her know that she will be humiliated until she does.

Diane Abbott has toughed it out for more than three decades. But many other women give up. In November 2019, several of the eighteen female members of Parliament who announced that they would not seek reelection the following month reported that the vicious abuse was a key factor in the decision.

"I am exhausted by the invasion into my privacy and the nastiness and intimidation that has become commonplace," MP Heidi Allen wrote in a letter to her constituents explaining why she was returning to private life. "Nobody in any job should have to put up with threats, aggressive emails, being shouted at in the street, sworn at on social media, nor have to install panic alarms at home."

In 2019, MP Paula Sherriff requested in the House of Commons that Prime Minister Boris Johnson tone down his "offensive, dangerous, inflammatory" language, which often resulted in threats of abuse to mostly female politicians. She spoke passionately about all the death threats she and her colleagues received, threats that quoted the prime minister. *You fat bitch. Stupid cow. I won't be happy till you're hanging from a lamppost.* Johnson, however, dismissed the abuse as "humbug," accusing politicians of creating the contentious political climate that caused the abuse.

Similarly, Michigan governor Gretchen Whitmer called on President Donald Trump to stop inciting violence against her. Each time he derided her at a rally, the number of threats she received skyrocketed. Rage at her went further than mere online abuse. In October 2020, a group of thirteen Trump-supporting militia members were arrested for plotting to kidnap Whitmer, try her for treason, and possibly execute her.

Instead of dialing down his rhetoric, only days later Trump attacked her at a rally for closing the schools during the worst months of the coronavirus pandemic. "Lock her up," the crowd chanted. "Lock her up," Trump repeated, smiling. "Lock 'em *all* up."

Whitmer tweeted that Trump's attack was "exactly the rhetoric that has put me, my family, and other government officials' lives in danger. . . . It needs to stop."

At one 2016 election event, Trump seemed to indicate Hillary Clinton should be shot. "Hillary wants to abolish—essentially abolish the Second Amendment [on gun rights]. By the way, if she gets to pick her judges, nothing you can do, folks," Trump told a North Carolina rally. "Although the Second Amendment people—maybe there is, I don't know."

On November 7, 2021, Arizona representative Paul Gosar tweeted a ninety-second, photoshopped, anime-style video of him cutting the throat of New York representative Alexandria Ocasio-Cortez. If Gosar had been a member of the public, Twitter would have removed the video. But as public officials are apparently allowed to tweet violence, Twitter permitted the post to remain, calling it "in the public interest," though it did slap a warning label on it, stating that it violated the rules on "hateful conduct." The post caused a national outcry. Representative Ted Lieu of California tweeted, "In any workplace in America, if a coworker made an anime video killing another coworker, that person would be fired."

Gosar removed the tweet but indicated his critics were making a mountain out of a molehill, that they clearly had no sense of humor at all. His opponents were making "a gross mischaracterization of a short anime video." He stated that his tweet "was not meant to depict any harm or violence against anyone portrayed" (which makes us wonder if he believes slitting someone's throat is harmless and nonviolent). He claimed the video was "a symbolic portrayal of a fight over immigration policy. . . . No matter how the left tries to quiet me I will speak out against amnesty for illegal aliens."

At no time did Gosar apologize to Representative Ocasio-Cortez. In a November 17 speech on the House floor before a vote on censuring Gosar, Minority Leader Kevin McCarthy called the move for censure "an abuse of power," pointed to the bad behavior of Democratic members (though not one case involved threats of murder to their colleagues), and ranted about inflation and high gas prices, among many other Democratic crimes. Addressing the chamber soon after, Ocasio-Cortez said, "It is a sad day in which a member who leads a political party in the United States of America cannot bring themselves to say that issuing a depiction of murdering a member of Congress is wrong and instead decides to venture off into a tangent about gas prices and inflation. What is so hard? What is so hard about saying that this is wrong?"

The House censured Gosar mostly along party lines, with just two Republicans (Liz Cheney of Wyoming and Adam Kinzinger of Illinois) joining in. Apparently, all the other Republicans got the video's hilarious joke. Minutes after the censure, Gosar defiantly retweeted the post.

Laura Boldrini, Speaker of the Italian Parliament from 2013 to 2018, has been threatened with gang rape and decapitation and has been burned in effigy. One day she received a bullet in the mail. "Death to Boldrini" has been spray-painted on countless walls all over Italy. She stays in a safe house while running for election. "The ones that hate migrants and the ones that hate women in positions of power—it's the same cultural framework," she explained. She took to posting the names of her abusers on her Facebook page.

A 2016 Inter-Parliamentary Union study on sexism, harassment, and violence against women parliamentarians surveyed female MPs from thirty-nine countries. More than 80 percent of the respondents said they had experienced abuse, and 44 percent had received threats of murder, rape, brutality, or the kidnapping and murder of their children. One European MP had received more than five hundred rape threats on Twitter over a period of four days. Oddly, some abusers threaten the women with *not* raping them because they are too ugly to be raped (not likely to be a threat to instill fear and horror in their victims). Some 20 percent reported that they had been slapped, pushed, and punched.

"SORRY IF ANYONE WAS OFFENDED"

A province in the heart of Canada, Alberta saw an explosion of sexist abuse after Rachel Notley became premier in 2015. "They're not calling her an idiot, they're calling her the c-word," gender consultant Cristina Stasia told the Canadian Broadcasting Corporation. "They're not saying she's too progressive, they're calling her a bitch. And there's a fury that lurks underneath this about the fact that we have a woman running our province."

Male Albertan politicians had never experienced the vitriol leveled at Notley. One survey called Notley the most threatened Albertan premier ever. Alberta, a province with a cowboy history now focused mainly on the oil and gas industry, did not take kindly to a female premier seeking to address climate change.

"Someone's gotta man up and kill her," posted one hater. Another said, "That dumb bitch is going to get herself shot." Other posts suggested Notley be shot, stabbed, and thrown into a tree grinder. One meme featured a photo of Notley as seen through a rifle's scope. When Notley appointed the first gender-balanced cabinet in Canadian history, the sexist fury exploded in outrage again.

The organizers of a golf tournament for oil executives erected a large picture of Notley as a target for participants to try to hit with their golf balls. A video appeared online showing two men laughing as they ran over the photo with their golf cart. The message was clear: Notley deserved to be whacked hard with golf balls. She deserved to be run over. When contacted about the misogynistic behavior, the organizer lamely said he was "sorry if anyone was offended."

When in 2016 a candidate for leadership of an opposing political party, Chris Alexander, held a rally, the chant "Lock her up!" reverberated across the crowds. Trump's rant, it seemed, had become international.

What is going on psychologically here? Is there some Jungian archetype of the evil feminine deeply rooted in the human subconscious that we want to kill or, in our slightly more civilized era, throw in jail? Are men afraid of being unmanned? As Fox News host Tucker Carlson put it, if Hillary Clinton were president, "How long do you think it would take before she castrates you?"

The result of "Off with her head" is often the successful silencing of women. A 2014 Australian study found that most women who had considered a career in politics were less likely to pursue one due to all the misogyny thrown at Julia Gillard. In 2020, Blair Williams, who wrote her PhD dissertation on the media coverage of five female prime ministers in the UK, Australia, and New Zealand, told the Australian Broadcasting Company, "You see a lot more girls and women who are saying they don't really want to enter politics, because they don't want to experience that kind of sexism, they don't want to have their entire personal lives up for critique."

The end result of threats and abuse is that women and girls inclined to pursue politics may decide it simply isn't worth it. Who would want to deal on an almost daily basis with threats of rape, death, and the kidnapping of their children? In silencing them, the Misogynist's Handbook puts them in their place. The Patriarchy wins.

CHAPTER 15

RIPPING UP THE MISOGYNIST'S HANDBOOK

Your silence will not protect you.

—Poet and feminist Audre Lorde, 1934–1992

What would the world look like if misogyny was consigned to the garbage bin of history? If roughly half the world's presidents, prime ministers, senators, MPs, governors, mayors, and other politicians were women? In an interview for this book, Francesca Donner, former gender director of the *New York Times*, initially had difficulty imagining such a world. "An emoji with its brain exploding," she finally replied. "It would be a radically different place. Imagine if the Fortune 500 companies had 250 female CEOs. Or the US Congress with 50 percent women. What legislation would be prioritized? Bills regarding childcare, healthcare, the struggles of working women? We would start hearing from families about what mattered. . . . Imagine if our hope for girls and boys was the same."

Is such a world even possible? After thousands of years of beheading, skewering, silencing, and shaming politically powerful women, is misogyny ever really going to go away? What can we do, here and now, with all the tools we have at hand, to put an end to such a dominant mindset adversely affecting half the human population? How do we purge ourselves of the powerful paleomisogyny clinging to us through hundreds of generations like an everlasting curse?

One way to at least curtail the abomination is to force social media platforms to prevent the rapid proliferation of false and sexist information about women politicians. On August 6, 2020, more than a hundred female politicians around the world—including House Speaker Nancy Pelosi and Representatives Jackie Speier, Alexandria Ocasio-Cortez, and Ilhan Omar—signed a letter to Facebook CEO Mark Zuckerberg and COO Sheryl Sandberg insisting that the company take steps to fight sexism on its platform, particularly that lobbed at female political candidates. "Much of the most hateful content directed at women on Facebook is amplified by your algorithms," the letter stated, "which reward extreme and dangerous points of view with greater reach and visibility creating a fertile breeding ground for bias to grow."

A few days earlier, on July 30, Facebook chose not to remove an altered video of Nancy Pelosi in which she appeared to be falling-down drunk (she has repeatedly stated that she never touches alcohol) by digitally slowing down her speaking. While Twitter and YouTube removed the fake video, Facebook slapped a "partially true" label on it. (Which part is true? The fact that she's alive?) Within days, some 2.6 million people had watched it.

A Facebook spokesperson replied to the letter by email, stating that the firm was working on the problems mentioned "in a variety of ways," which included "technology that identifies and removes potentially abusive content before it happens, by enforcing strict policies, and by talking with experts to ensure we stay ahead of new tactics." A response that is about as unspecific as it is unconvincing.

In an interview for this book, Lucina Di Meco, cofounder of the #ShePersisted Global Initiative, said, "There is a reality that unless we change the way social media platforms work, it is going to be very hard in the long term for a balanced discourse to be heard. In reality, the way platforms are designed is to encourage outrageous content because it generates more engagement whether it is truthful or not; the more outrageous, the more engagement. Also, social media companies have not been very good at keeping up with their own terms of service, eliminating certain bad actors, eliminating misinformation. They have promised but have not really done so. There is only so much that women and their supporters can do to balance negative discourse when it is so pervasive, and the women might not even see it."

Di Meco pointed out that social media platforms have a lack of accountability imposed upon almost every other industry. "When a company makes a cheese and sells it in supermarkets," she said, "we are not trusting the cheesemaker. Someone is going to inspect the company, the machines, the supply chain, and then decide if the cheese is ready to go to market or if eating that cheese would have negative consequences. Social media companies started as a small thing, and nobody thought that there was any reason to regulate them. But now we know they have had an impact on democracy, on mental health, on the riots on the Capitol, so actually somebody needs to regulate them. They have not had a positive track record."

Representative Jackie Speier, who co-chairs the Democratic Women's Caucus, blames social media for the increase in threats and violence against women lawmakers. In an interview with *Recode* in August 2020, she said, "There have been so many threats on my life over the length of my service. Two in the last two years actually were taken up by the local district attorney, and individuals were convicted. So there has been an increase of this kind of vile behavior, and it's got to stop. And we're putting Facebook on notice that they've got to be part of bringing some normalcy back to this process and to espouse their mission about diversity and inclusion and empowerment."

In March 2020, the presumptive Democratic nominee Joe Biden announced that he would choose a woman as his vice presidential nominee, and as the months passed, pressure was on him to select a Black woman. Many women's organizations sprang into action to form the Women's Disinformation Defense Project, which would spend more than $20 million on ads, research, and strategies to stop racist and sexist tropes online as they occurred. They were well prepared when, on August 11, Biden did indeed choose a Black woman. The organizations identified sexist images of Kamala Harris on Facebook and Twitter—riding a broomstick, snakes wriggling in her hair, and all those awful sexualized memes—and called on the platforms to remove them.

NARAL, a nonprofit organization that advocates for expanded

access to abortion and birth control, "deputized" some of their organization's 2.5 million members as spokespeople to call out sexism on social networks when they saw it. "What we know is when it comes to voters, the best surrogates are the people in their own communities who they respect," its president Ilyse Hogue told the *Washington Post* a few days before the announcement.

TIME'S UP Now, a charity that raises money to support victims of sexual harassment, created a nonpartisan "SWAT team" to defend women politicians against sexist attacks and go after those responsible. "Whenever this subtle, and not so subtle, bias creeps into public discourse, we will fight back and shine a light on it before it takes hold," the organization stated on its website just hours before Biden made his announcement. "We will share this information widely with allies as part of an unprecedented effort to shift the narrative about women running for office, once and for all." Tina Tchen, president and CEO, added, "When our politics focus on a woman's likeability or ambition instead of her experience and expertise, we all lose out. We will not allow these attacks, which have stamped out the political ambitions of countless qualified women and kept others from pursuing office in the first place, to go unanswered."

Just as important as social media is the role of the traditional news networks, websites, and other publications in portraying candidates and elected leaders by describing and critiquing them. The women's groups were poised to tackle sexism in the mainstream media as well.

Shortly before Biden's announcement of Kamala Harris as his

pick, the group We Have Her Back sent a letter to the newsroom leaders of the top media in the country. Signed by NARAL's Ilyse Hogue, EMILY's List president Stephanie Shriock, and many other top executives of national women's organizations, the letter asked news media to consider carefully how they would represent Biden's pick and other women in the upcoming election. "There are multiple ways that media coverage over the years has contributed to the facts of the lack of diversity at the top of society's roles," the letter stated, and mentioned several, including, "Reporting on a woman's ambition as though the very nature of seeking political office, or any higher job for that matter is not a mission of ambition. Reporting on whether a woman is liked (a subjective metric at best) as though it is news when the 'likeability' of men is never considered a legitimate news [story]. Reporting, even as asides in a story, on a woman's looks, weight, tone of voice, attractiveness and hair is sexist news coverage unless the same analysis is applied to every candidate."

The letter concluded, "We believe it is your job to, not just pay attention to these stereotypes, but to actively work to be anti-racist and antisexist in your coverage (ie: equal) as this political season progresses and this Presidential ticket is introduced. As much as you have the public's trust, you also have great power. We urge you to use it wisely."

UltraViolet, too, put out media guidelines on how to avoid sexist tropes when reporting on female candidates. Titled "Reporting in an Era of Disinformation: Fairness Guide for Covering Women and People of Color in Politics," it points out that a candidate must be evaluated on her experience, her past decisions, and her ability to step into the top job. The guide then asks reporters and commentators:

- Are you punishing women and celebrating men for doing the same thing?
- Are you suggesting ambition is a bad thing?
- Are you putting too much emphasis on appearance?
- Are you focusing on her tone of voice—shrill, bitter, angry—rather than the substance of her statements?
- Are you analyzing or focusing on her clothing?
- Are you focusing on weight loss or gain?
- Are you focusing on her makeup and hair?
- Are you telling a candidate to smile or talking about whether she smiles?
- Are you hypersexualizing a candidate or politician?
- Are you commenting on her attractiveness?
- Are you using words like "unlikeable" or "unelectable"?
- Are you questioning her commitment to the United States based on the color of her skin or country of origin?
- Are you calling a Black woman angry?

The guide asked that the press, when reporting on a sexist or racist social media post, not publish it—which would only exponentially increase its coverage—but merely describe it as sexually or racially offensive.

"We are putting the media on notice," UltraViolet's executive director, Shaunna Thomas, told CBS News on August 15, 2020.

"We are not going to allow the proliferation of racist and sexist coverage of these women to dominate the headlines and to impact the way voters understand them."

Not all of the groups' efforts had the desired effect. On October 6, 2020, TIME'S UP Now released a report that found significant sexist and racist media coverage of Kamala Harris's selection. One quarter of all coverage featured tropes such as the "angry black woman" and the she-doesn't-belong-here "birther" falsehood. The report also found that attacks on Harris were far more vicious than those on Hillary Clinton's VP pick, Senator Tim Kaine, or Trump's choice, Governor Mike Pence, in 2016. Harris's opponents tore into her with the sexist tropes of "nasty," "phony," and "mean," while white men Kaine and Pence had been deemed a tad dull.

Lucina Di Meco of #ShePersisted, however, saw improvement in much of the mainstream media coverage of Harris. "Happily, we seem to have learned a lot from the experience of 2016," she said. "I think it was very different in the traditional media coverage. Journalists were a lot more prepared not to replicate the harmful posts."

MORE DIVERSE NEWSROOMS

Another path to reducing misogyny in the press is to create more diverse newsrooms, especially among senior editors. According to a spring 2018 article in the *Columbia Journalism Review*, 90 percent of the top editors at the 135 most widely distributed newspapers were white, and 73 percent were male, statistics that usually influence how women and people of color are covered. Women of color made up just 7.9 percent of traditional newspaper staff, 12.6 percent of local TV news staff, and 6.2 percent of local radio news staff, according to the Women's Media Center's report "The Status of Women of Color in the U.S. News Media 2018."

Some progress has been made recently. Sally Buzbee was named executive editor of the *Washington Post* in May 2021, the first woman in the position. A week earlier, the *Los Angeles Times* appointed Kevin Merida, who is Black, as editor. And the *New York Times* has had a Black editor, Dean Baquet, since 2014. Nicole Carroll has been editor in chief of *USA Today* since 2014. Of course, women and people of color in positions of journalistic power don't necessarily tamp down sexism and racism. Fox News has had a female CEO since 2018 and has, if anything, continued to spew out even more misogynistic vitriol. But, generally speaking, people from diverse backgrounds have less tolerance of isms than those from white male monoculture, which is reflected in their reporting.

PREPARING WOMEN CANDIDATES FOR THE ONSLAUGHT

Most new candidates, male and female, undergo training in public speaking, dealing with the media, campaign strategy, and other subjects. These days, women candidates should also receive training in how to deal with the sexism slung by journalists and trolls alike. In an interview for the study "Women, Politics & Power in the New Media World," Liz Grossman, cofounder and CEO at social impact firm Baobab Consulting, said, "In order for women to best harness the media, they should hire experts who can prepare them for interviews and questioning, who can also enforce strict rules with journalists regarding which questions they can ask, and which subjects are taboo."

Lucina Di Meco feels that women candidates should also take training on how to defend themselves against the psychological pain of online abuse. "When they understand that the abuse isn't personal, it is helpful," she said. "That it wasn't their haircut or the dress they wore or the thing they said that ruined them. That it was going to happen regardless of the dress or haircut. That it's systemic."

"SHAME THE SHAMERS"

When Julia Gillard became prime minister of Australia in 2010, she couldn't help but notice the vicious sexism lobbed against her— how could she not, what with it splayed across the Internet, the major newspapers and networks? But how to respond to it? She decided not to be accused of "playing the woman card," looking like a whiner. Nor did she want to dignify such silliness with a response. As Gillard wrote in her autobiography, she decided "to tolerate all the sexist and gendered references and stereotyping on the basis it was likely to swirl around for a while and then peter out. I was wrong," she admitted. "It actually worsened. Should I have been clearer about it all earlier? Started press conferences by taking to task particularly stupid sexism in reporting? Would it have made a difference or only started allegations of playing the gender wars earlier? Honestly, I do not know."

In her book *Women and Leadership*, she wrote, "What I found was the longer I served as Prime Minister, the more shrill the sexism became. Inevitably governments have to make tough decisions that some people like and others hate. That is certainly true of the government I led. What was different was that the go-to weapon in hard political debates became the kind of insults that only get hurled at a woman. That emerged as a trend alongside what was already a highly gendered lens for viewing my prime ministership. Every negative stereotype you can imagine—bitch, witch, slut, fat, ugly, child-hating, menopausal—all played out."

In a July 2020 TV interview, she said, "I do muse to myself that, you know, the second day I was prime minister, the news media was entirely about the jacket I wore. Like, no one reported anything I said the second day I was prime minister. It was all about what I was wearing. And I wonder now if, you know, on the third day I was prime minister, if I'd gone out to the Canberra press pack and said, 'Is anybody feeling a little bit silly about this? If I'd been a bloke wearing a suit, would you have put that on the news yesterday? "Oh, my God, he's got a charcoal suit on!" Would anybody cover that? Are we going to keep doing this for as long as I'm prime minister?'"

In 2016, Hillary Clinton, too, decided not to call out the misogyny. She swatted it away gently as if it were a vaguely disturbing gnat. As a result, it swelled in both its viciousness and its deadly efficacy. "We all wanted to believe in 2016 that a woman could run on her own qualifications," said NARAL's Ilyse Hogue in 2020, "and we found out that's not true. . . . We will take nothing for granted this time around."

In her 2021 International Women's Day video, Namibian first lady Monica Geingos discussed the gendered insults she had been subjected to—stupid, unqualified, too ambitious, fat, ugly, and slut—and how she finally decided to respond to them. Many online abusers, she noted, blame all the country's failings on her, though she wields no power. "When I am not busy being a manipulative, deceitful gold digger," she said, "I am busy running the country as I bewitched my old sugar-daddy husband who is too blind to see through my feminine charms. . . . I mean, surely you can see that the president is a good man. The problem—wait for it—the problem is his wife. The only reasonable explanation for poor government decisions is that he has been influenced by his corrupt, greedy, interfering, controlling, horrible wife." (Just like Eve, Jezebel, Cleopatra, Anne Boleyn, and Marie Antoinette.)

At first, Geingos took the threats and insults with patriarchal-approved silence. "When there is a clear social media campaign of anonymous Whatsapp messages specifically targeting me in the most disgusting ways, I was told not to respond, to ignore them," she said. "And I did. It was a mistake. I was wrong." She quoted the words of Caribbean American feminist poet Audre Lorde, "'Your silence will not protect you.' The insults just got worse, and the lies they were willing to tell became increasingly outrageous. There were no more boundaries. My parents, my children, my family, my friends, all my loved ones became targets."

Geingos decided to fight back. "Power doesn't concede without a demand and neither does patriarchy." She said she recently instituted a defamation lawsuit against a particularly vicious troll. "An interesting thing happens when you stand up for yourself, when you challenge," she said. "You'll be called a troublemaker, too aggressive, too unladylike. That is why many of us prefer not to challenge gender bias. That is why we ignore being called gold diggers, sluts, Delilahs. That is why you will ignore being told you are too fat, too thin, your clothes are too tight, and you should not have an opinion on politics. . . . I will not be silenced anymore. . . . If I allow myself to be silenced, bullied, and insulted, I may be signaling that this conduct is okay, that it's normal. . . . It's not okay, it's not normal. . . ."

But how can women call out sexism without looking like whiners? In an interview for this book, Amanda Hunter of the Barbara Lee Family Foundation advised women to link their response to a "larger belief system," as Representative Alexandria Ocasio-Cortez so deftly did in her landmark speech in the US Capitol in July 2020. Rather than complaining that Representative Ted Yoho insulted her personally by calling her a "fucking bitch," Ocasio-Cortez put him and other misogynists on the defensive by calling sexism a "cultural" problem, where men "accost women without remorse and with a sense of impunity." "This issue is not about one incident," the congresswoman said.

Hunter explained the speech "was so powerful because AOC did not focus on why her feelings were hurt. Voters don't care if an individual woman's feelings are hurt. AOC focused on why sexist behavior is harmful to all women and girls."

Hunter also pointed to Kamala Harris's handling of Vice President Pence's frequent interruptions during their October 2020 debate as the most effective way for a female politician to deal with misogyny. As he kept rolling over her, Harris firmly said, "Mr. Vice

President, I'm speaking." "For voters, it's really a leadership test," Hunter said. "Harris had a calm demeanor and tone of voice, and yet she held her ground. That is the tightrope voters want women to walk."

Lucina Di Meco said, "We see more and more from research, that when women respond to sexist attacks, it benefits them. Women were told to fly high, not to respond or address the sexism. We now know that those strategies didn't work; they made the attackers louder and made the women seem weak."

Di Meco added that when social media users see another woman abused, they should use the platform to denounce the troll and support the victim. "For women in politics, in particular, it is crucial they are not the only ones to give that response. It needs to come strongly from their support networks so the positive voices outweigh or balance the negative voices. The power of the pack, of the network, is something positive that can be used as a source of hope."

Di Meco recommended that women, "Call out sexism, denounce online harassment, and respond to negative ads. Shame the shamers. Build a community of online supporters who will pile onto the aggressors when they post harassing comments on social media. Don't leave the field of battle to the enemy. Get out there and fight back! Drown out the misogyny."

In an interview for "Women, Politics & Power in the New Media World," Italian congresswoman Laura Boldrini, former president of the Chamber of Deputies, pointed out the countless online supporters who fought back against her abusers. "Social media is truly a double-edged sword," she said. "I became a target of politically motivated, vicious online attacks carried out by armies of trolls who used sexism and fake news, trying to silence and delegitimize me. Yet, as I exposed and denounced the trolls and harassers, thousands of people came to my defense online, claiming the digital space as an arena to denounce sexism and shape the political discourse."

Helle Thorning-Schmidt, prime minister of Denmark from 2011 to 2015, told *Time* in 2020, "If bad behavior doesn't have a consequence, then it becomes a lesson to anyone who wants to behave badly that they can just carry on with impunity. So I think that there comes a time where enough is enough."

GET MEN INVOLVED

Men—journalists, politicians, commentators, CEOs, and community leaders—need to stand up for women in the face of sexism. Julia Gillard has often speculated about what a difference it would have made if, while gendered critiques of her prime ministership were being hurled around, a leading Australian man from outside politics had been prepared to say publicly, "As Australians we do not do our politics this way. Let's have a political debate that is respectful and free of gender stereotyping."

"If I had that time again," she wrote, "I would reach out to community leaders beyond the world of politics, men in particular, and try to get them involved in calling out the sexism. These voices would have been seen as more objective than my own." On the Australian TV program *Q&A*, she said, "I think if the CEOs of Australia's top 10 leading companies—the day after the rally with the 'bitch witch' signs—if they'd done a letter to the newspaper which said, 'Look, people can have a variety of views about putting a price on carbon, they're all legitimate views, we should be having a debate, but we don't have a debate calling the prime minister of the country with sexist terms,' I think that would have been really noted."

GET MORE WOMEN ELECTED

One way to reduce misogyny in political office is to elect so many women that they no longer seem to be oddities. Amanda Hunter described the "imagination barrier" that hurts women candidates, the fact that many voters have difficulty picturing "women in power, an area dominated by white men." She said, "Seeing more women in office, seeing a woman vice president, chips away at the imagination barrier and takes down stereotypes. Women governors often open the door to other women governors. We will change the conversation by having more women run for and get elected to office."

Similarly, journalist Christina Cauterucci wrote in *Slate* in November 2019, "The only thing that will inoculate the public to the jarring novelty of women in positions of power is more women in positions of power. The waning of sexism in politics won't be marked by people starting to *like* women in leadership but by the decline of likability as a political criterion—by people not liking female candidates, the same way they don't like male candidates, and voting for them anyway."

Ursula von der Leyen, president of the European Commission, said that elected women should support other women in politics, as

German chancellor Angela Merkel did by nominating her as a cabinet minister. "My experience is that women tend to be hesitant, not to grab the opportunity too easily," von der Leyen said in a podcast for International Women's Day in 2021, "and it's our responsibility, mainly as female leaders, to encourage them and to tell them 'I believe in you, I'm sure you can do it and I'll support you.'"

IS THE HANDBOOK'S POWER FADING?

It has been a long journey from Hatshepsut being chiseled off temple walls to the sexist and racist memes of Kamala Harris. Eve and the apple. Pandora and the box. Jezebel's eye makeup. Catherine de Medici's Saint Bartholomew's Day Massacre. Catherine the Great and the horse. Julia Gillard's empty fruit bowl. Hillary's child sex ring under the pizza parlor.

In all that time, women in positions of power have been caricatured as untrustworthy trollops and vile vixens, hormonally enraged shrews and backstabbing bitches. As monstrous beings: witches, harpies, Furies, and snake-haired Medusas. More recently, they have also been depicted as shrill women with bad hair and the wrong clothes who play the gender card, too ambitious and selfish to take care of their husbands and children.

The clarion calls to put a nonconforming woman in her place resound more loudly now than ever before, what with the Internet and social media. Lock her up. Send her back. Off with her head. Even more so if she's Black or Brown. The axe falls. The cell door clangs shut, and the key turns in the lock. She loses the election. She loses her head. Will things change? *Can* they change?

If they can, now is the time. Even as social media threatens and abuses women and people of color, so does it create powerful movements that help us move in the direction of long-awaited equity. We live in an age of increasing questioning of the status quo. Of #MeToo, Black Lives Matter, and the removal from positions of prominence of statues of Confederate traitors who shed American blood to keep human beings enslaved.

Due to social media and the Internet, we are growing quite aware of the tried-and-true tactics of the Misogynist's Handbook. Those things that we may not have thought twice about ten or twenty years ago stick out at us like sore thumbs. A court declaring Britney Spears crazy for thirteen years as her father forced her to perform, controlled the hundreds of millions of dollars she was clearly sane enough to earn, and, as she claimed, forced her to keep an IUD in her uterus suddenly seems outrageous to us. And we all know such long-lasting brutality would never have happened to her for so many years if her name had been Brian Spears.

And, these days, criticizing women politicians for their hair, their shoes, the circumference of their hips, their voice, and their personal lives is just as likely to shower the critic with opprobrium as the target. The war room strategies of the women's organizations during elections are also bound to have some effect with the more progressive media, especially those with diverse newsrooms.

Perhaps, though, as with so many things, the clearest path to a better future lies with young people. Amanda Hunter said, "I am hopeful for the future, and I think things are changing. Look at the next generation, at young children. Those kids are never going to remember a time that there were not multiple women and people of color on the debate stage and running for president. During COVID-19, women governors and mayors have been on the national stage. Look at the little girls dressing up as Kamala Harris for Halloween. It's really powerful when you think about the fact that when I was their age, I did not have role models in elected office. That right there is a sign of change."

In her final speech as Australia's prime minister, Julia Gillard sounded a note of optimism for the women who will, one day, follow in her footsteps. She said, "What I am absolutely confident of is it will be easier for the next woman, and the woman after that, and the woman after that."

Naturally, there is furious backlash against this long-delayed crawl toward justice, resulting in the increasing viciousness of sexist tweets and memes, more violent threats, more savage abuse. Perhaps this sound and fury is less alarming when we realize that we are hearing the bellowing of a gravely wounded dinosaur, who has ruled the earth uncontested for tens of thousands of years, suddenly understanding the possibility of its own extinction, and raging against the dying of the light.

ACKNOWLEDGMENTS

To my editor at William Morrow, Lucia Macro, many thanks for your support through so many books. We have come a long way since you were my first US book editor back in 2003 for *Sex with Kings*. It is always such a pleasure to work with you, and I have enjoyed laughing with you about the strange foibles of life, liberty, and the pursuit of publishing.

To my agent Stephen Barbara at InkWell Management, I cherish your calm, reliable manner and incisive literary expertise. We have some weird and wonderful adventures on this journey. Here's to many more!

I have had a tremendous support group who kept me upbeat in the bizarre time of COVID: my ever-patient husband, Michael (though I sometimes feel I should call him "Job") Dyment; my stepsons, Phil and Sam Dyment; their SOs, Crisy Dyment and Marissa Goldberg; and my fabulous coterie of friends: Susanne Spencer, Elise Kress, Leslie Harris, Adrienne Harris, Shirley Ginwright, Helen Randolph, King Peggy, Emily Heddleson, Kelly

Ghaisar, and Tim Kime. Special thanks to Rev. Sylvia Sumter, Senior Minister of Unity of Washington, DC, for the joyful light of her wisdom. Thanks to all of you for buoying me up, laughing at my jokes, and keeping me from sliding into irreversible pandemic brain fog.

My friends' love, laughter, and support have been particularly helpful during the pandemic. A time when, instead of lecturing at book events, and going to TV and radio stations for interviews, and meeting fans, and signing their books, I do all my publicity on Zoom with my butt in the exact same chair where I have been sitting for two years writing. Sometimes I feel like a fly stuck on sticky paper. *Help! I am stuck in this chair, and I can't get up!* I so miss live events where the air sizzles and crackles with excitement and expectation, where energy becomes synergy, and ideas are exchanged, and emotions spark, and every one of us in that room contributes to creating something completely new.

I sincerely hope, with this book, things will be different, and I will get to meet many of you (and get my butt out of this chair). Fingers crossed!

BIBLIOGRAPHY

VIDEOS YOU ABSOLUTELY MUST WATCH

Hillary Clinton told NBC News that "the more a woman is in service to someone else": https://www.youtube.com/watch?v=4r4XG-IytVw

Australian prime minister Julia Gillard's Misogyny Speech: https://www.youtube.com/watch?v=fCNuPcf8L00

European Commission president Ursula von der Leyen denied a seat by President Erdoğan of Turkey: https://www.youtube.com/watch?v=4OEm_gUi8gY

Ursula von der Leyen's reaction to Sofagate: https://www.youtube.com/watch?v=YPd6dn9mkaw

Representative Alexandria Ocasio-Cortez responds to Representative Ted Yoho calling her a bitch: https://www.youtube.com/watch?v=tSZcdMCHn2o

Representative Alexandria Ocasio-Cortez responds to Representative Paul Gosar's photoshopped anime tweet in which he kills her: https://www.youtube.com/watch?v=wcCD3RI2vMg

Maxine Waters stands up for women in Congress: https://www.c-span.org/video/?59115-1/whitewater-controversy-house-floor

Kamala Harris grills Attorney General Bill Barr: https://www.youtube.com/watch?v=wcvcj_VuywA

Trevor Noah lambastes those who say Kamala Harris isn't Black enough: https://www.thedailybeast.com/trevor-noah-tears-into-fox-news-for -saying-kamala-harris-isnt-black-enough

BOOKS

Adams, Tracy. *The Life and Afterlife of Isabeau of Bavaria.* Baltimore: The Johns Hopkins University Press, 2010.

Aitken, Jonathan. *Margaret Thatcher: Power and Personality.* New York: Bloomsbury, 2013.

Andrews-Dyer, Helena, and R. Eric Thomas. *Reclaiming Her Time: The Power of Maxine Waters.* New York: Dey Street, 2020.

Anonymous. *A Mervaylous Discourse upon the Lyfe, Deeds, and Behaviours of Katherine de Medicis, Queen Mother: wherein are displayed the meanes which she had practised to attain unto the usurping of the Kingedome of France, and to the bringing of the estate of the same unto utter ruine and destruction.* Publisher unknown. Heidelberg, 1575.

Asimov, Isaac. *Guide to the Bible.* New York: Wings Books, 1968.

Ball, Molly. *Pelosi.* New York: Henry Holt & Co., 2020.

Beard, Mary. *Women & Power: A Manifesto.* London: Profile Books, 2018.

Beauvoir, Simone de. *The Second Sex.* New York: Vintage Books, 1989.

Bhutto, Benazir. *Daughter of Destiny: An Autobiography.* New York: Simon & Schuster, 1989.

Bordo, Susan. *The Creation of Anne Boleyn: In Search of the Tudors' Most Notorious Queen.* London: Oneworld Publications, 2018.

Bordo, Susan. *The Destruction of Hillary Clinton.* Brooklyn: Melville House, 2017.

Bordo, Susan. *Imagine Bernie Sanders as a Woman.* Denver: Outskirts Press, 2020.

Bourdeille, Pierre de, abbé de Brantôme, *Book of the Illustrious Dames.* Freiburg, Germany: Outlook Verlag, 2020.

Bradford, Sarah. *Lucrezia Borgia*. New York: Viking, 2004.

Bridge, Antony. *Theodora: Portrait in a Byzantine Landscape*. Chicago: Academy Chicago Publishers, 1993.

Bulfinch, Thomas. *Bulfinch's Mythology*. New York: Viking Press, 1979.

Bunce, Robin, and Samara Linton. *Diane Abbott: The Authorised Biography*. London: Biteback Publishing, 2020.

Burkett, Elinor. *Golda Meir & The Birth of Israel*. London: Gibbon Square, 2017.

Chisholm, Shirley. *Unbought & Unbossed*. Boston: Houghton Mifflin, 1970.

Clinton, Hillary Rodham. *What Happened*. New York: Simon & Schuster, 2017.

Cook, Blanche Wiesen. *Eleanor Roosevelt, Volume 1: The Early Years, 1884–1933*. New York: Penguin, 1992.

Cooney, Kara. *When Women Ruled the World: Six Queens of Egypt*. Washington, DC: National Geographic, 2018.

Cooper, Helene. *Madame President: The Extraordinary Journey of Ellen Johnson Sirleaf*. New York: Simon & Schuster, 2017.

Crankshaw, Edward. *Maria Theresa*. New York: Viking Press, 1969.

de Hart, Jane Sherron. *Ruth Bader Ginsburg: A Life*. New York: Alfred A. Knopf, 2018.

Dolan, Julie, Melissa Deckman, and Michele L. Swers. *Women and Politics: Paths to Power and Political Influence*. Boston: Longman, 2011.

Doyle, Sady. *Dead Blondes and Bad Mothers: Monstrosity, Patriarchy, and the Fear of Female Power*. Brooklyn: Melville House, 2019.

Doyle, Sady. *Trainwreck: The Women We Love to Hate, Mock, and Fear . . . and Why*. Brooklyn: Melville House, 2016.

Enloe, Cynthia. *The Curious Feminist*. Berkeley: University of California Press, 2004.

Erickson, Carolly. *Great Catherine: The Life of Catherine the Great, Empress of Russia*. New York: St. Martin's Griffin, 1994.

Ferraro, Geraldine. *My Story*. New York: Bantam Books, 1985.

Fitzpatrick, Ellen. *The Highest Glass Ceiling: Women's Quest for the American Presidency*. Cambridge, Massachusetts: Harvard University Press, 2016.

Frank, Katherine. *Indira: The Life of Indira Nehru Gandhi*. New York: Houghton Mifflin, 2002.

Fraser, Antonia. *Marie Antoinette: The Journey*. New York: Anchor Books, 2001.

Friedan, Betty. *The Feminine Mystique*. New York: W. W. Norton & Company, 1997.

Gillard, Julia. *My Story*. Sydney, Australia: Vintage Books, 2015.

Gillard, Julia, and Ngozi Okonjo-Iweala. *Women and Leadership: Real Lives, Real Lessons*. Sydney, Australia: Penguin Random House Australia, 2020.

Gordon, Mary. *Joan of Arc*. New York: Viking Press, 2000.

Grant, Michael. *Cleopatra: A Biography*. New York: Barnes & Noble Books, 1995.

Han, Lori Cox, and Caroline Heldman, editors. *Rethinking Madam President*. Boulder, Colorado: Lynne Rienner Publishing, 2007.

Holland, Jack. *A Brief History of Misogyny: The World's Oldest Prejudice*. London: Robinson, 2006.

Jamieson, Kathleen Hall. *Beyond the Double Bind: Women and Leadership*. Oxford: Oxford University Press, 1995.

Jansen, Sharon L., trans. *Anne of France: Lessons for My Daughter*. Cambridge: D. S. Brewer, 2004.

Kingdon, Robert M. *Myths about the St. Bartholomew's Day Massacres, 1572–1576*. Cambridge, MA: Harvard University Press, 1988.

Knecht, R. J. *Catherine de' Medici*. London: Longman, 1988.

Knox, John. *The First Blast of the Trumpet Against the Monstrous Regiment of Women*. Geneva: Jean Crespin, 1558.

Komisar, Lucy. *Corazon Aquino: The Story of a Revolution*. New York: George Braziller, 1987.

Krook, Mona Lena. *Violence Against Women in Politics*. Oxford: Oxford University Press, 2020.

Lampitt, Dinah. *The King's Women.* New York: Signet Books, 1992.

Lowry, Joan A. *Pat Schroeder: A Woman of the House.* Albuquerque: University of New Mexico Press, 2003.

Mahoney, Irene. *Madam Catherine.* New York: Coward, McGann & Georghegan, Inc., 1975.

Manne, Kate. *Down Girl: The Logic of Misogyny.* Oxford: Oxford University Press, 2018.

Manne, Kate. *Entitled: How Male Privilege Hurts Women.* New York: Crown, 2020.

Morain, Dan. *Kamala's Way: An American Life.* New York: Simon & Schuster, 2021.

Murray, Rainbow, ed. *Cracking the Highest Glass Ceiling: A Global Comparison of Women's Campaigns for Executive Office.* Santa Barbara, CA: Praeger, 2010.

Nolan, Hayley. *Anne Boleyn: 500 Years of Lies.* New York: Little A, 2019.

Paglia, Emile. *Sexual Personae: Art and Decadence from Nefertiti to Emily Dickinson.* New York: Vintage, 1991.

Phillips, Jess. *Everywoman: One Woman's Truth about Speaking the Truth.* London: Windmill Books, 2017.

Plowden, Alison. *Marriage with My Kingdom: The Courtships of Queen Elizabeth I.* New York: Stein and Day, 1977.

Prince, Rosa. *Theresa May: The Enigmatic Prime Minister.* London: Biteback Publishing, 2017.

Procopius. *The Secret History.* New York: Penguin, 1966.

Qvortrup, Matthew. *Angela Merkel: Europe's Most Influential Leader.* New York: Overlook Duckworth, 2017.

Reeves, Rachel. *Women of Westminster: The MPs Who Changed Politics.* London: I.B. Taurus, 2019.

Sandberg, Sheryl. *Lean In: Women, Work, and the Will to Lead.* New York: Alfred A. Knopf, 2013.

Schiff, Stacy. *Cleopatra: A Life.* New York: Little, Brown & Co., 2010.

Schroeder, Pat. *24 Years of House Work . . . and the Place Is Still a Mess: My Life in Politics*. Kansas City: Andrew McNeel Publishing, 1998.

Sloan, Susan. *A Seat at the Table: Women, Diplomacy, and Lessons for the World*. Potomac, MD: New Degree Press, 2020.

Starkey, David. *Six Wives: The Queens of Henry VIII*. New York: Harper-Collins, 2003.

Steinberg, Blema S. *Women in Power: The Personalities and Leadership Styles of Indira Gandhi, Golda Meir, and Margaret Thatcher*. Montreal: McGill Queen's University Press, 2008.

Suvorova, Anna. *Benazir Bhutto: A Multidimensional Portrait*. Oxford: Oxford University Press, 2015.

Thomas, Chantal. *The Wicked Queen: The Origins of the Myth of Marie Antoinette*. New York: Zone Books, 2001.

Thomas, Evan. *First: Sandra Day O'Connor, An Intimate Portrait of the First Woman Supreme Court Justice*. New York: Random House, 2019.

Trimble, Linda. *Ms. Prime Minister: Gender, Media, and Leadership*. Toronto: University of Toronto Press, 2017.

Tyldesley, Joyce. *Hatshepsut: The Female Pharaoh*. New York: Penguin Books, 1998.

Verdon, Jean. *Isabeau de Bavière*. Paris: J. Tallendier, 1981.

Weir, Alison. *The Life of Elizabeth I*. New York: Ballantine Books, 1998.

Williamson, Hugh Ross. *Catherine de' Medici*. London: Michael Joseph, Ltd., 1973.

WEB ARTICLES AND SCHOLARLY PAPERS

Aalberg, Toril, and Anders Todal Jenssen. "Gender Stereotyping of Political Candidates, An Experimental Study of Political Communication." *Nordicom Review* 28, no. 1 (2007): 17–32.

Abend, Lisa. "'The Idea of Denmark as a Gender Paradise Is a Myth.'

#MeToo Finally Came to Denmark—And the Reckoning Is Only Just Beginning." *Time*, November 2, 2020.

Adams, Tim. "Testosterone and High Finance Do Not Mix: So Bring on the Women." *Guardian*, June 11, 2011.

Adams, Tracy, and Glenn Rechtschaffen. "The Reputation of the Queen and Public Opinion: The Case of Isabeau of Bavaria." *Medieval Feminist Forum* 47, no. 1 (2011).

Akutekha, Esther. "What Kamala Harris' Style Choices Say about Her Politics." *HuffPost*, October 7, 2020.

Allen, Jonathan. "Confessions of a Clinton Reporter: The Media's 5 Unspoken Rules for Covering Hillary." *Vox*, July 6, 2015.

Allsop, John, Kelsey Ables, and Denise Southwood. "Who's the Boss?" *Columbia Journalism Review* (Spring/Summer 2018).

Atir, Stav, and Melissa J. Ferguson. "How Gender Determines the Way We Speak about Professionals." *Proceedings of the National Academy of Sciences of the United States* 115, no. 28 (July 2018): 7278–7283.

Atwell, Ashleigh. "Mayor of Virginia Town Refuses to Resign, Defends Meme of Aunt Jemima as Biden's VP: 'I Took It to Be Humorous.'" *Atlanta Black Star*, August 6, 2020.

Bigio, Jamille. "Women, Peace, and Security." https://www.cfr.org /conference-calls/women-peace-and-security.

Boboltz, Sarah. "Jill Biden Spokesman Snaps Back at *Wall Street Journal* for Editorial on 'Dr.' Title." *HuffPost*, December 12, 2020.

Burnett, Sara. "Democrats See Racism in GOP Mispronunciations of 'Kamala.'" Associated Press, August 22, 2020.

Cadelego, Christopher, and Natasha Korecki. "Karen Bass Rises as a Sleeper Pick to be Biden's VP." *Politico.com*. July 24, 2020.

Cardy, Meghan. "Lock Her Up: Harassment and Violence Against Women in Alberta Politics." *Political Science Undergraduate Review* 3, no. 1 (January 2017).

Carlin, Diana B., and Kelly L. Winfrey. "Have You Come a Long Way, Baby? Hillary Clinton, Sarah Palin, and Sexism in 2008 Campaign Coverage." *Communication Studies* 60, no. 4 (September–October 2009).

Carr, Glynda C. "A Black Woman Was Speaking and the World Listened." *The Crisis Magazine*, October 10, 2020.

Catanese, David. "The Authenticity Test." *USNews.com*. January 10, 2019.

Chemaly, Soraya. "Bernie Sanders Isn't Sexist. The Whole System Is." *Washington Post*, January 23, 2020.

Cheng, William. "The Long, Sexist History of 'Shrill' Women." *Time*, March 23, 2016.

Cooper, Marianne. "For Women Leaders, Likability and Success Hardly Go Hand-in-Hand." *Harvard Business Review*, April 30, 2013.

Devere, Heather. "The Don and Helen New Zealand Election 2005: A Media A-gender?" *Pacific Journalism Review*, December 2006.

Di Meco, Lucina. "Women, Politics & Power in the New Media World." *She-Persisted.org*, November 5, 2019.

Doyle, Sady. "Why Do Successful Women Like Hillary Clinton Get Under So Many People's Skin?" *Guardian*, November 6, 2016.

Evans, Jon, and Jerel Slaughter. "Gender and the Evaluation of Humor at Work." *Journal of Applied Psychology* 104, no. 8 (February 7, 2019): 1077–1087.

Fang, Marina. "Like Clockwork, Kamala Harris Criticized as 'Unlikeable' Because She's a Woman in Politics." *HuffPost*, October 8, 2020.

Frenkel, Dean. "Drop the Gillard Twang: It's Beginning to Annoy." *Sydney Morning Herald*, April 21, 2011.

Friedman, Avery. "Michelle Obama Addresses the Racism She Faced as First Lady." *Global Citizen*, July 26, 2017.

Friedman, Vanessa. "Boris Johnson and the Rise of Silly Style." *New York Times*, July 23, 2019.

Friedman, Vanessa. "Why Covering Nancy Pelosi's Hot Pink Dress Isn't Sexist." *New York Times*, January 10, 2019.

Gaines, Janet Howe. "How Bad Was Jezebel?" *Bible Review*, October 2000.

Gaines, Janet Howe. "Lilith: Seductress, Heroine, or Murderer?" *Bible Review*, October 2001.

Gibbons, Rachel. "Isabeau of Bavaria, Queen of France (1385–1422): The Creation of an Historical Villainess." *Transactions of the Royal Historical Society* 6 (1996): 51–73.

Givhan, Robin. "Hillary Clinton's Tentative Dip Into New Neckline Territory." *Washington Post*, July 20, 2007.

Goldberg, Emma. "Fake Nudes and Real Threats: How Online Abuse Holds Back Women in Politics." *New York Times*, June 3, 2021.

Gothreau, Claire. "The Objectification of Women in Politics and Why It Matters." *Center for American Women in Politics*, August 31, 2020.

Gray, Emma. "'F**king Bitch' and the Everyday Terror Men Feel about Powerful Women." *HuffPost*, July 23, 2020.

Guerin, Cecile, and Eisha Maharasingam-Shah. "Public Figures, Public Rage: Candidate Abuse on Social Media." *The Institute for Strategic Dialogue*, October 5, 2020.

Gupta, Boshika. "Here's What Made Cleopatra So Different from Other Egyptian Rulers." *Grunge.com*, September 16, 2020.

Hall, Lauren, J., and Ngaire Donaghue. "'Nice Girls Don't Carry Knives': Constructions of Ambition in Media Coverage of Australia's First Female Prime Minister." *British Journal of Social Psychology* 52, no. 4 (August 2012).

Hampton, Janie. "The Taboo of Menstruation." *Aeon.com*, May 2, 2017.

Hao, Karen. "Deepfakes Have Got Congress Panicking. This Is What It Needs to Do." *MIT Technology Review*, June 12, 2019.

Harris, John F., and Peter Baker. "Terry Ties Allen to Attacks." *Washington Post*, October 13, 1993.

Harvey, Josephine. "'Like A Boss': Twitter Users Are Raving about Kamala Harris' Campaign Trail Shoes." *HuffPost*, September 8, 2020.

Hattersley, Giles. "'I'm Just Trying To Constantly Remind People, 'Don't Forget Women' — Labour MP Jess Phillips." *British Vogue*, August 2019.

Heilweil, Rebecca. "What Women in Congress Want from Facebook." *Vox*, August 7, 2020.

Hesse, Monica. "The *Wall Street Journal* Column about Jill Biden Is Worse Than You Thought." *Washington Post*, December 13, 2020.

Hosack, Kristen. "Can One Believe the Ancient Sources That Describe Messalina?" *Constructing the Past* 12, no. 1 (2011).

Howey, Catherine L. "Dressing a Virgin Queen: Court Women, Dress, and Fashioning the Image of England's Queen Elizabeth I." *Early Modern Women* 4 (Fall 2009): 201–208.

Howze, Candace. "If You Think Calling Us 'Angry Black Women' Is an Insult, You'd Better Think Again." *HuffPost*, August 21, 2020.

Janes, Chelsea. "Kamala Harris Could Be Quietly on the Brink of a Historic Leap." *Washington Post*, November 1, 2020.

Kendall, Mikki. "22 Times Michelle Obama Endured Rude, Racist, Sexist or Plain Ridiculous Attacks." *Washington Post*, November 16, 2016.

Kim, Eun Kyung. "Obama Apologizes to Kamala Harris for 'Best-looking Attorney General' Comment." *Today.com*, April 5, 2013.

Korecki, Natasha, Christopher Cadelago, and Marc Caputo. "She had no Remorse. Why Kamala Harris isn't a Lock for VP." *Politico.com*, July 27, 2020.

Kranish, Michael. "In Her First Race, Kamala Harris Learned How to Become a Political Brawler." *Washington Post*, October 5, 2020.

Krook, Mona Lena. "How Sexist Abuse of Women in Congress Amounts to Political Violence—and Undermines American Democracy." *The Conversation*, October 21, 2020.

Leader, Lauren. "The Role We ALL Play in Keeping Sexism Out of Presidential Politics." *NBC News.com*, August 11, 2020.

Lester, Amelia. "Ladylike: Julia Gillard's Misogyny Speech." *New Yorker*, October 9, 2012.

Lewis, Michael. "Obama's Way." *Vanity Fair*, September 11, 2012.

Linskey, Annie, and Isaac Stanley-Becker. "Biden Campaign, Women's Groups Are Working to Blunt Sexist Attacks on His Vice Presidential Pick." *Washington Post*, August 8, 2020.

Manns, Keydra. "Michigan Deputy Fired after Sharing Racist Photo of Kamala Harris Watermelon Jack-O'-Lantern." *Yahoo.com*, November 19, 2020.

Marcus, Ruth. "The Politics of Hair Is Fraught. And Hell Hath No Fury Like a Woman Filmed at Her Salon." *Washington Post*, September 4, 2020.

Martin, Jonathan, Alexander Burns, and Katie Glueck. "Lobbying Intensifies Among V.P. Candidates as Biden's Search Nears an End." *New York Times*, August 10, 2020.

Mazza, Ed. "Trump Slammed after Firing Off Outrageous New Insults at Kamala Harris." *HuffPost UK*, September 9, 2020.

McCarthy, Ellen. "For Female Leaders, Humor Is a Blessing. Unless It's a Curse." *Washington Post*, January 16, 2020.

McIlvanna, Una. "'A Stable of Whores?' The 'Flying Squadron' of Catherine de Medici," in *The Politics of Female Households: Ladies-in-waiting across Early Modern Europe*, eds. Nadine Akkerman and Brigid Houben. Leiden: Brill, 2014.

Milligan, Susan. "Women Candidates Still Tagged as Too 'Emotional' to Hold Office." *USNews.com*, April 16, 2019.

Moore, Martha T. "Focus on Hillary Clinton's Appearance Sparks Criticism." *USA Today*, May 9, 2012.

Murray, Rainbow. "Nicola Sturgeon Has Triumphed over Britain's Sexist Press—and That's Good for All Women." *The Conversation*, May 6, 2015.

Newton-Small, Jay. "How Hillary Clinton Is Trying to Avoid Being 'Shrill.'" *Time.com*, February 3, 2016.

Okimoto, Tyler G., and Victoria L. Brescoll. "The Price of Power: Power

Seeking and Backlash Against Female Politicians." *Personality and Social Psychology Bulletin*, June 2, 2010.

Papenfuss, Mary. "Michigan Gov. Lashes Trump for Endangering Her Life after Rally 'Lock Her Up' Chant." *HuffPost*, October 17, 2020.

Pardes, Arielle. "The Evolution of the Bitch." *Vice.com*, September 9, 2014.

Paresys, Isabelle. "Dressing the Queen at the French Renaissance Court: Sartorial Politics," in *Sartorial Politics in Early Modern Europe*, ed. Erin Griffey. Amsterdam: Amsterdam University Press, 2019.

Petri, Alexandra. "Team Biden Wishes It Had Known Something Was Slightly Wrong with Every Woman Before Promising a Female VP." *Washington Post*, August 11, 2020.

Plott, Elaina. "Why Donald Trump's Hateful Rhetoric Against Women Is Working." *Harper's Bazaar*, July 27, 2016.

Reese, Diana. "News Coverage of Female Candidate's Appearance Damages Her Chance of Winning." *She The People*, April 8, 2013.

Reeves, Rachel. "Power Dressing: Why Female MPs Have Faced a Century of Scrutiny." *Guardian*, March 2, 2019.

Rucker, Philip, and Isaac Stanley-Becker. "Trump and Allies Struggle to Find a Focused Attack on Harris." *Washington Post*, August 12, 2020.

Ryzik, Melena, Reggie Ugwu, Maya Phillips, and Julia Jacobs. "When Trump Calls a Black Woman 'Angry,' He Feeds This Racist Trope." *New York Times*, August 14, 2020.

Schaberg, Jane. "How Mary Magdalene Became a Whore." *Bible Review*, October 1992.

Scott, Eugene. "In Accusations of Being Too Ambitious, Some Black Women See a Double Standard." *The Fix*, August 3, 2020.

Seo, Diane. "Reigning Women: 'Year of the Woman'? In Central L.A. They've Wielded Political Clout for Years—and More Are on the Way." *Los Angeles Times*, November 29, 1992.

Sheridan, Chris. "Why Bernie Sanders Won't Brush His Hair." *Al Jazeera*, November 19, 2015.

Sheridan, Mary Beth. "Mexico's Bold Break with Machismo: Congress Is Now Half Female, and Gender Parity Is the Law." *Washington Post*, September 9, 2021.

Smith, David. "We Still Have a Problem with Female Authority: How Politics Sets a Trap for American Women." *Guardian*, March 6, 2020.

Solender, Andrew. "Trump Calls Harris 'Mad Woman' as Top Campaign Advisor Questions Her VP Eligibility." *Forbes*, August 13, 2020.

Specia, Megan. "Threats and Abuse Prompt Female Lawmakers to Leave U.K. Parliament." *New York Times*, November 1, 2019.

Spring, Mariana, and Lucy Webster. "A Web of Abuse: How the Far Right Disproportionately Targets Female Politicians." *BBC Newsnight*, July 15, 2019.

Sreberny-Mohammadi, Annabelle, and Karen Ross. "Women MPs and the Media: Representing the Body Politic." *Parliamentary Affairs* 49, issue 1 (January 1996): 103–115.

Stevis-Gridneff, Matina, and Carlotta Gall. "Two Presidents Visited Turkey. Only the Man Was Offered a Chair." *New York Times*, April 7, 2021.

Stewart, Emily. "The Attacks on Ilhan Omar Reveal a Disturbing Truth about Racism in America." *Vox*, December 4, 2019.

Sullivan, Margaret. "Tucker Carlson's Mangling of Kamala Harris's Name Was All About Disrespect." *Washington Post*, August 12, 2020.

Sullivan, Margaret. "With Biden Likely to Pick a Black Woman as VP, Here's How the Media Can Avoid Playing into Sexist and Racist Tropes." *Washington Post*, August 10, 2020.

Sullivan, Patricia. "Local Virginia GOP Leader Posts Sexist Meme about Kamala Harris on Party's Online Page." *Washington Post*, August 14, 2020.

Sutherland, N. M. "Catherine de Medici: The Legend of the Wicked Italian Queen." *The Sixteenth Century Journal* 9, no. 2 (July 1978).

Taylor-Smither, Larissa J. "Elizabeth I: A Psychological Profile." *The Sixteenth Century Journal* 15, no. 1 (Spring 1984).

Traister, Rebecca. "'It Was No Accident.' Congresswoman Pramila Jayapal on Surviving the Siege." *The Cut*, January 8, 2021.

Traister, Rebecca. "The Ugly Truth About the RNC: Donald Trump is Not an Outlier." *New York* magazine, July 20, 2016.

Tumulty, Karen, Kate Woodsome, and Sergio Peçanha. "How Sexist, Racist Attacks on Kamala Harris Have Spread Online—a Case Study." *Timesupnow.org*, October 7, 2020.

Uscinski, Joseph E., and Lilly J. Goren. "What's in a Name? Coverage of Senator Hillary Clinton during the 2008 Democratic Primary." *Political Research Quarterly*, published online September 22, 2010.

Vartan, Starre. "Why Do Men Tell Women to Smile?" *Mnn.com*, September 17, 2017.

Verveer, Melanne, and Lucina Di Meco. "Gendered Disinformation, Democracy, and the Need for a New Digital Social Contract." *The Council Foreign Relations*, May 6, 2021.

Vogelstein, Rachel B., and Alexandra Bro. "Women's Power Index," Council on Foreign Relations, March 21, 2021.

Walsh, Joan. "In 2020, Double Standards Are Still Dogging Elizabeth Warren—and All the Women Candidates." *The Nation*, January 21, 2020.

Williams, Blair. "From Tightrope to Gendered Trope: A Comparative Study of the Print Mediation of Women Prime Ministers." Doctoral dissertation, Australian National University, School of Politics and National Relations, September 19, 2019.

Woodward, Marian. "Ditch the Witch: Julia Gillard and Gender in Australian Public Discourse." Thesis, Department of Gender and Cultural Studies, University of Sydney, 2013.

Worth, Anna, Augoustinos Martha, Brianne Hastie. "'Playing the Gender Card': Media Representations of Julia Gillard's Sexism and Misogyny Speech." *Feminism & Psychology* 26, issue 1 (October 2015): 52–72.

Yakas, Ben. "Hot-And-Bothered *NY Post* Columnist Slut-Shames Danish Prime Minister." www.gothamist.com, December 12, 2013.

Zhouli, Li. "Use of the Word 'Bitch' Surged after Women's Suffrage." *Vox*, August 19, 2020.

Zitser, Joshua. "Trump Supporter Furious after Being Banned from Fox News for 'Misogynistic' Attack on Kamala Harris." *Indy100.com*, October 9, 2020.

Zoonen, Liesbet van. "The Personal, the Political and the Popular: A Woman's Guide to Celebrity Politics." *European Journal of Cultural Studies* 9, issue 3 (August 2006): 287–301.

NO AUTHOR NAMED, WEBSITES

"Fairmont Councilman Retracts Use of 'Hoe' from Harris Statement." *WDTV*, August 13, 2020.

"Intimidation in Public Life: A Review by the Committee on Standards in Public Life Presented to Parliament by the Prime Minister by Command of Her Majesty." https://www.gov.uk/government/collections /intimidation-in-public-life (December 2017).

"Media Guide to Gender Neutral Coverage of Women Candidates + Politicians." NameItChangeIt.org, 2012.

"Reporting in an Era of Disinformation: Fairness Guide for Covering Women and People of Color in Politics." UltraViolet, 2020.

"Staying Power: Strategies for Women Incumbents." The Barbara Lee Family Foundation, June 2021.

"The Status of Women of Color in the U.S. News Media 2018." The Women's Media Center, March 6, 2018.

"Time's Up on Sexism in Politics." Timesupnow.org, August 7, 2020.

"Trudeau Family's Attire Too Flashy Even for an Indian?" *Outlook India*, February 21, 2018.

"Von der Leyen sollte zunächst nicht mit aufs Foto." *Frankfurter Allgemeine Zeitung*, April 8, 2021.

"When Women Run." FiveThirtyEight.com, 2020.

"Women's Participation in Peace Processes." Council on Foreign Relations, cfr.org, 2019.

SPEECHES

Summers, Anne, Ph.D. "The Political Persecution of Australia's First Female Prime Minister (Vanilla Version)." 2012 Human Rights and Social Justice Lecture, University of Newcastle, August 31, 2012.